Scientific Essentialis

Scientific Essentialism defends the view that the fundamental laws of nature depend on the essential properties of the things on which they are said to operate and are therefore not independent of them. These laws are not imposed on the world by God, the forces of nature, or anything else, but rather are immanent in the world.

Brian Ellis argues that ours is a dynamic world consisting of more or less transient objects that are constantly interacting with each other and whose identities depend on their roles in these processes. Natural objects must behave as they do because to do otherwise would be contrary to their natures. The laws of nature are therefore metaphysically necessary, and consequently there are necessary connections between events.

In an innovative contribution to contemporary metaphysics, Ellis calls for the rejection of the theory of Humean Supervenience and implementation of a new kind of realism in philosophical analysis.

This book will interest professionals and students of philosophy and the philosophy of science.

Brian Ellis is Professor Emeritus of Philosophy at La Trobe University, Victoria, Australia, and Professorial Fellow in the Department of History and Philosophy of Science at the University of Melbourne. His books include *Basic Concepts of Measurement* (Cambridge University Press, 1966), *Rational Belief Systems,* and *Truth and Objectivity.*

CAMBRIDGE STUDIES IN PHILOSOPHY

General editor ERNEST SOSA (Brown University)

Advisory editors:
JONATHAN DANCY (University of Reading)
JOHN HALDANE (University of St. Andrews)
GILBERT HARMAN (Princeton University)
FRANK JACKSON (Australian National University)
WILLIAM G. LYCAN (University of North Carolina at Chapel Hill)
SYDNEY SHOEMAKER (Cornell University)
JUDITH J. THOMSON (Massachusetts Institute of Technology)

Scientific Essentialism

BRIAN ELLIS

La Trobe University and University of Melbourne

CAMBRIDGE
UNIVERSITY PRESS

CAMBRIDGE UNIVERSITY PRESS
Cambridge, New York, Melbourne, Madrid, Cape Town, Singapore, São Paulo

Cambridge University Press
The Edinburgh Building, Cambridge CB2 8RU, UK

Published in the United States of America by Cambridge University Press, New York

www.cambridge.org
Information on this title: www.cambridge.org/9780521800945

First published 2001
This digitally printed version 2007

A catalogue record for this publication is available from the British Library

Library of Congress Cataloguing in Publication data
Ellis, B. D. (Brian David), 1929–
Scientific essentialism / Brian Ellis.
p. cm. – (Cambridge studies in philosophy)
ISBN 0-521-80094-3
1. Essence (Philosophy) 2. Science – Philosophy. I. Title. II. Series.
B105.E65 E45 2001
149–dc21   00-063094

ISBN 978-0-521-80094-5 hardback
ISBN 978-0-521-03774-7 paperback

To Scamp

Who kept vigil on the floor behind my chair
To guard me from whatever threats there might be
From hostile forces
And defenders of Humean Supervenience

Contents

Preface

This book had its origin in a series of discussions I had with John Bigelow and Caroline Lierse, beginning in 1989, on the idea that the world is an instance of a natural kind and that the laws of nature are of its essence. This is not a new idea. Rom Harré and E.H. Madden (1975) had expressed a similar view in their excellent book, *Causal Powers*. But we thought that the idea was worth a more thorough investigation than it had had up to that time, and the three of us wrote a joint paper on this topic (Bigelow, Ellis, and Lierse, 1992).

At the time, I was thinking about the problem of objective knowledge in the particular form in which it arose out of the work that I was then doing on *Truth and Objectivity*. If the world has no intrinsic structure, I thought, then, in principle, any single way of conceptualizing the world might be as good as any other. There would be an objective world, and perhaps from a human perspective, a best description of it. But such a description might not also be the best from the point of view of an alien being, whose epistemic values might well be different from ours. However, if the world had a natural kind structure, I reasoned, then we should be able to classify things in the world objectively in two quite different ways – vertically, in terms of the distinct objects that are the members of the natural kinds, and horizontally – in terms of the kinds to which they belong, and hence the properties they must have by virtue of their memberships of these kinds. In that case, there would be a set of objective facts about the world – facts that would exist independently of how we, or any other being, might think or reason about the world. The hypothesis of a natural-kind structure of reality thus promised to yield a solution to the objective knowledge problem, for objective knowledge could then be defined with reference to this structure.

For three years, from 1990, Caroline Lierse worked with me as a research assistant, and then in 1993 as a research associate. In this period, we wrote a series of papers on a range of topics, including dispositional properties, laws of nature, and the ontology of natural kinds. Joint papers were read at national conferences of the Australasian Association of Philosophy and Social Studies of Science in 1992 and 1993. One of these was later published as "Dispositional Essentialism." This book is heavily dependent on the work we did together in this period, and I am greatly indebted to Caroline for her contribution to it. I am particularly indebted for her work on powers and dispositions (in Chapter 3), for her arguments against Humeanism (published separately in Lierse, 1996), for contributions, too numerous to detail, to other sections of the work, and for the encouragement and support she gave me in pursuit of the project.

The book has also profited from discussions with John Bigelow, John Fox, David Armstrong, Alan Chalmers, Keith Campbell, George Molnar, George Bealer, and Erik Anderson, and with several overseas correspondents, including Evan Fales, Chris Swoyer, and Storrs McCall. I wish to thank them all for their contributions, and pay special tribute to the members of the School of Philosophy at La Trobe University and the Department of History and Philosophy of Science at the University of Melbourne, who put up with my going on and on about natural kinds, laws, and essential natures, and doing it all so very goodnaturedly.

For long periods of time, my wife, Jenny, has also had a lot to put up with, with my sitting in front of my computer (even when it was a beautiful day outside and I should have been out in the garden). I thank her for her tolerance and support, without which this project would never have been completed.

Introduction

There are two very different theories about how the laws of nature relate to the world. One is that they are somehow imposed on things whose identities are independent of the laws. The other is that the laws are immanent in the world, not superimposed on it. On the second theory, the laws of nature depend on the essential properties of the things on which they are said to operate, and are therefore not independent of them.

The first is the dominant world-view of modern philosophy. It is a view that was shared by Descartes, Newton, Locke, Hume, and Kant – the founding fathers of all of the major philosophical traditions of Western Europe. This view is closely associated with seventeenth- and eighteenth-century mechanism, and like mechanism, it implies that the laws of nature are contingent and operate on things that are essentially inert or passive. If things were naturally active, then they would be bound to act according to their natures, rather than as the imposed laws of nature might require. Consequently, the dominant world-view implies that what happens in the world depends essentially on what the laws of nature happen to be. It does not depend only on what kinds of things there are in the world, or in what circumstances they exist.

The second world-view is essentialism. It is the sort of theory I will be defending in this book. It is not a view that has been widely accepted in modern times. One has to go all the way back to Aristotle to find a truly notable defender of essentialism. Yet, essentialism is precisely the sort of theory that one would expect any modern scientific realist to accept. For a realist would now be hard pressed to make much sense of the passive, and intrinsically inert, world on which the laws of nature are supposed to operate. The world, according to modern science, seems not to be innately passive, but fundamentally active and reactive. It is certainly not a mecha-

1

nistic world of things having only the attributes of extension and impenetrability, as Descartes' and Locke's worlds were. Rather, it is a dynamic world consisting of more or less transient objects which are constantly interacting with each other, and whose identities would appear to depend on their roles in these processes.

I will assume that this appearance is also reality. Thus, I will assume that it is impossible – metaphysically impossible – for a proton or any other fundamental particle to have a causal role different from the one it actually has. The assumption is plausible, I suggest, because a proton would appear to have no identity at all apart from its role in causal processes. If this is right, then the laws concerning the behavior of protons and their interactions cannot be just accidental – that is, laws which could well have been otherwise. On the contrary, it is essential to the nature of a proton that it be disposed to interact with other things as it does. Its causal powers, capacities, and propensities are not just accidental properties of protons, which depend on what the laws of nature happen to be, but essential properties, without which there would be no protons, and which protons could not lose without ceasing to exist (or gain without coming into being). The idea that the laws of particle physics are superimposed on intrinsically passive things which have identities that are independent of the laws of their behavior thus lies very uneasily with modern science.

Scientific essentialism is proposed as a metaphysic for scientific realism which is compatible with this intuition, and also with the evident dynamism of modern science. It is not a reversion to Aristotelianism, or an attempt to resuscitate medieval views about the nature of reality. On the contrary, its origins are decidedly twentieth century. It depends, for example, on the existence of quantum discreteness in the world. For it is this discreteness at the quantum level which ultimately guarantees that there are real, ontologically based distinctions of kind in nature. That is, the discreteness of quantum reality is the generator of the real distinctions between natural kinds of substances, properties, and processes. And this fact is crucial for the viability of modern essentialism. Aristotle was able to believe in natural kinds (although he did not call them this) because of the diversity and apparent distinctness of the various animal and plant species that he knew about. But the idea that biological species are natural kinds has not survived criticism, especially in light of Darwin's theory of evolution. The new essentialism, however, is not a defense of Aristotle's theory of *biological* kinds, and does not depend on it. It depends rather on the recognition that there are hierarchies of natural *physical* kinds, formally like

2

the supposed hierarchies of biological species, but real kinds, nevertheless, based on quantum discreteness, and existing at a deeper level.

The distinctions between the chemical elements, for example, are real and absolute. There is no continuum of elementary chemical variety which we must arbitrarily divide somehow into chemical elements. The distinctions between the elements are there for us to discover, and are guaranteed by the limited variety of quantum mechanically possible atomic nuclei. Many of the distinctions between kinds of physical and chemical processes are also real and absolute. There is no continuum of processes within which the process of β-emission occurs, and from which it must be arbitrarily distinguished. The world is just not like that. At a fundamental level, the processes that occur often allow real and absolute distinctions of kind to be made. Therefore, if there are natural kinds of objects or substances, as I believe there are, there are also natural kinds of events and processes.

Scientific essentialism is thus concerned with natural kinds, which range over events or processes, as well as with the more traditional sort, which range only over objects or substances. The natural kinds in these two categories evidently occur in natural hierarchies. At the apex of the hierarchies, I postulate that there are two very general natural kinds. The most general natural kind in the category of substances includes every other natural kind of substance existing in the world. This is the global kind, for our world, in the category of substances. It defines the range of natural kinds of objects or substances that can exist in our world. The most general kind in the category of events is the global kind, which includes every other natural kind of event or process which occurs in the world. This global kind effectively defines the range of kinds of events or processes that can occur in our world.

At the bases of these two hierarchies are the infimic species of objects and events. These are species that have no sub-species, and whose members are therefore essentially identical. To illustrate: The most fundamental particles that we know about are plausibly members of infimic species in the hierarchy of natural kinds of substances. In the hierarchy of natural kinds of events, the infimic species are plausibly just quantitatively identical natural kinds of events or processes. These infimic species have all of the formal characteristics of classical universals, except that they range over different kinds of things. I call them *substantive* and *dynamic* universals, respectively, in order to distinguish them from each other, and also from the more usual *property* universals, whose instances are tropes, or property instances.

3

I postulate that the world itself is a member of a natural kind that is distinguished from worlds of all other kinds by the global natural kinds of objects or events, to which it is host, and by its fundamental ontology of objects, events, and properties. If this is right, then the most general laws of nature can be seen to arise from the essential properties of these most general natural kinds of objects and events. The more specific laws of nature – the causal laws and the laws defining the structures of the natural kinds of substances that exist in the world – might then be seen as deriving from the essential properties of the more specific kinds of substances and events occurring in the world. According to scientific essentialism, therefore, all of the laws of nature, from the most general (for example, the conservation laws and the global structural principles) to the more specific (for example, laws defining the structures of molecules of various kinds, or specific laws of chemical interaction) derive from the essential properties of the object and events that constitute it, and must hold in any world of the same natural kind as ours.

Scientific essentialism challenges orthodoxy in philosophy in a number of ways. It requires us to believe that there are ontologically irreducible causal powers in nature – dispositional properties that define the causal roles of their bearers. In the standard Humean view, dispositional properties are all ontologically dependent on the categorical (non-dispositional) properties of things, and on what the laws of nature happen to be. They cannot therefore be fundamental properties, but must supervene on an underlying categorical reality. But, according to scientific essentialism, there are causal powers in nature, and some of them are ontologically primary. They do not depend on what the causal laws of nature happen to be. On the contrary, the causal laws just describe the natural kinds of processes involved in their display.

Scientific essentialism also requires us to reject the thesis known as Humean Supervenience – the thesis that all modal properties (for example, natural necessity, natural possibility, objective probability, and so on) supervene on non-modal properties. This thesis derives its plausibility from the Humean conception of reality as consisting of "atoms in the void" – of self-contained atomic objects or states of affairs that are located in space and time, and succeeding one another in ways determined by the laws of nature. The intuition to which this conception naturally gives rise is that if one can say which objects possessing what intrinsic qualities exist at which points in space and time, then one can describe the world completely. Therefore, any properties that describe what a thing must or might do in given circumstances must supervene on those that describe things as they are in themselves. That is, there could not be two worlds that are identical in respect of all non-modal properties that nevertheless differed in respect of modal properties.

If one accepts the Humean Supervenience thesis, then the problem of explaining what modal properties are, or what makes statements attributing such properties to things true, becomes acute. Some Humeans have gone to quite extraordinary lengths to accommodate them, interpreting all modal statements as claims about relationships holding between possible worlds. But since every possible world, according to Humeanism, is a world without modal properties, it is hard to see how this is supposed to solve the problem. A universe of worlds without modalities is a universe without causal powers. Scientific essentialists simply reject the Humean conception of reality, and the Supervenience Thesis that it entails. From the perspective of scientific essentialism, the world is not an agglomeration of logically independent states of affairs or self-contained atoms of any other kind. The world consists ultimately of things that have their causal powers essentially that determine what they can, must, or cannot do in relation to other things.

The causal powers of things are displayed in processes belonging to natural kinds. Processes of these kinds can only be brought about by appropriate activating events or circumstances, and what is essential to them as processes is precisely that they are the actions of such properties in such circumstances. There is therefore a close link between natural kinds of processes, causal powers, and the circumstances of their display. Specifically, if anything has a given causal power, then it must be disposed to act in the appropriate way in the appropriate circumstances. Moreover, if anything has this power essentially, then it is necessarily the case that it will be disposed to act in this way in circumstances of this kind.

It is one of the aims of physical science to discover and to describe the inherent causal powers of things. For these powers are the truth-makers for those laws of nature that are generally known as causal laws. Causal powers thus have a very important role in the world, for they are the sources of the immanent causal laws of nature.

The causal powers of an object are those of its properties that determine how it is disposed to behave. These properties are all essentially dispositional. That is, their identities depend on the dispositions they underpin. For any specific causal power, there is a range of possible displays, normally related by a probability function from the state of the object that possesses it, and the properties of the circumstances in which it can be activated. The range of possible ways in which a given causal power

5

can act then defines a natural kind of process. For there is no process, however similar in appearance, that is a process of this kind but is not an action or display of this causal power. Causal powers and the natural kinds of processes that are their displays thus go hand in hand, and are interdefinable.

The possible displays of any given causal power are all processes that are essentially similar in their structure and that differ from each other only in ways that lie within the permitted range variation for the kind of process concerned. The causal powers of an object are thus the real essences of the causal processes that can occur when that object acts causally. The gravitational mass of an object, for example, is the dispositional property it has that determines its causal role in generating gravitational fields, and hence the effects it has on other objects immersed in these fields. The charge on an object is the dispositional property it has that determines its causal role in generating electromagnetic fields, and hence the effects it has on other objects that are in or moving through these fields. Therefore, if anything, say a fundamental particle, has a certain mass and a certain charge essentially, then it must generate such fields in any world in which it might exist, and have precisely the same effects on things of just the same kinds.

The fact that the causal laws all ultimately depend on the causal powers of things belonging to natural kinds, and the manners and circumstances of their displays, has many implications. First, it implies that these laws of nature are metaphysically necessary. Since they are immanent in the world, the laws of nature cannot be changed, without the world itself being changed. And things of the kinds that do exist in this world could not exist in any other world in which the laws of nature affecting them are supposed to be different. Second, it implies that causal laws can only be found in areas concerned with natural kinds of objects, properties, or processes. Consequently, there are no laws of nature to be found in the social or human sciences, not even in economics. Third, it implies that the laws of nature are not just observable regularities. For natural processes are normally superimposed one on top of another, and to say anything sensible about any of them, they need to be separated, at least in thought. Therefore, the laws of nature, which are concerned with the description of natural kinds of processes, must often be expressed abstractly with reference to idealized objects in ideal circumstances.

This book will elaborate and defend a modern essentialist metaphysic, based on an ontology of natural kinds of objects, properties and processes.

This ontology will be developed in Chapter 2. As a metaphysic, it stands opposed to some widely held views about the nature of reality. These views mostly derive from the common seventeenth- and eighteenth-century belief in the essential passivity of matter. It is a view which is still widely held by English-speaking philosophers. Somewhat ahistorically, I call the metaphysic that derives from this belief, "Humeanism." Humeanism will not be the subject of any one chapter or section of the book, but will come under attack, in one way or another, throughout. For my aim is to present a comprehensive metaphysic that can be seen as an alternative to Humeanism.

As well as holding that matter is essentially passive, Humeans generally subscribe to the following theses:

1. That causal relations hold between logically independent events.
2. That the laws of nature are behavioral regularities of some kind that could, in principle, be found to exist in any field of inquiry.
3. That the laws of nature are contingent.
4. That the identities of objects are independent of the laws of nature.
5. That the dispositional properties of things are not genuinely occurrent properties – which would have to be the same in all possible worlds – but are somewhat phoney world-bound properties that depend on what the laws of nature happen to be.

Against these theses, I will argue for the view that nature is active, not passive, and that:

1. Causal relations are relations between events in causal processes. If an event of a natural kind that would activate a given causal power in a certain way occurs, an event of a natural kind which would then be an appropriate display of that power must also occur (even though the effect may sometimes be masked by other effects).
2. The laws of nature are not just behavioral regularities, although they imply the existence of underlying patterns of behavior, but descriptions of natural kinds of processes arising from the intrinsic properties of things belonging to natural kinds. There are, accordingly, no laws of nature in fields such as sociology or economics.
3. The laws of nature are not contingent, but metaphysically necessary. The same things in the states in which they currently exist would have to have the same behavioral dispositions in any world in which they might exist.
4. The identities of objects are not independent of the laws of nature. If the laws of nature were different, the things existing in the world would have to be different.

7

5. There are natural dispositional properties that are genuinely occurrent, and which therefore act in the same ways in all possible worlds. These include the causal powers of the most fundamental kinds of things, so that things of these same kinds, existing in any other world, would be disposed to behave in just the same ways.

In Chapter 1, I will develop some of the machinery that will be required for essentialist analysis, and elaborate on some of the themes to be developed in the book. The aim of this chapter is to set the stage for what is to follow. Most of the issues raised in this chapter will be taken up again and more thoroughly investigated in later chapters. Some of the concepts will no doubt be familiar to readers, and may be passed over fairly quickly. But others will not be. The causal concept of intrinsicality, for example, is non-standard. It is nevertheless one of the basic concepts of the essentialist framework, and it is one which science itself could not do without. The essential properties of things belonging to natural kinds are all intrinsic in the causal sense of intrinsicality, and they are the key to understanding what the truth-makers for the laws of nature are. They include such properties as causal powers, capacities, and propensities.

Chapter 2 describes the ontology that is suggested by the natural-kinds structure of the world. It is an ontology that includes hierarchies of natural kinds of objects, processes, and properties. It is not argued here that this hierarchy is irreducible. A more frugal ontology of some kind may be viable, or more useful in other ways. However, I am more concerned with explanatory adequacy than frugality. A good ontology should provide insight into, or offer some kind of explanation of, the salient general features of the world that have been revealed to us by science. One of these is surely the hierarchical structure of laws. Another is the hierarchical structures of natural kinds of objects, processes, and properties. One would therefore expect a good ontology to recognize the existence of these hierarchical structures, and a good metaphysic to link these two general features of reality in some way – which is precisely what scientific essentialism does.

Two important concepts required for the construction of any ontology are those of *ontological dependence* and *ontological reduction*. These concepts are developed and applied systematically to explain the relationships of ontological dependence between species in any given hierarchy. Interestingly, the required concepts imply that quantities and other generic universals are not ontologically dependent on their infimic species, as many have supposed must be the case, but rather the other way around. Specific quantitative universals are ontologically dependent on the generic quantitative universals of which they are species. Some, especially those who fail to distinguish clearly between infimic species and instances, may find this result surprising. However, while the existence of the genus plausibly depends on the existence of some instance of some infimic species, the existence of the genus does not depend on the existence of any particular infimic species. If this is right, it explains the primacy of quantities in laws of nature, and the relative unimportance of specific quantitative measures of things.

Chapter 3 begins the process of articulating a more active role for the things in the world in determining what the laws of nature are, and hence how things are naturally disposed to behave. The chapter contains a defense of the somewhat heretical view that there are genuinely occurrent dispositional properties in nature, as well as categorical ones. The arguments of the categorical realists and other Humeans that dispositional properties are world-bound, and depend on what the laws of nature happen to be in any particular world, are countered. Some dispositional properties, it is argued, are at least as fundamental as any categorical properties, and are plausibly essential properties of some of the things that have them. The thesis to be defended is one that we called "dispositional essentialism," and was defended in a paper (1994a) by Caroline Lierse and me with this title. There are genuine dispositional properties, we argued; and these are not properties that things have because of what the laws of nature are. Rather, the laws of nature are what they are, we contended, because things of various kinds have the dispositional properties they have essentially. That is, they would not be things of the kinds they are if they lacked these properties. Consequently, they would not be things of the kinds they are if they were not disposed to behave as the relevant laws require.

The view that things have causal powers of various kinds is not new to science. Such properties are presupposed in all causal explanations. It is of course possible that causal explanations might be reduced somehow to explanations of other kinds. They might, compatibly with Humeanism, be just covering-law explanations in disguise. But most scientists, I imagine, believe in causality, or would at least agree that some events may make other events objectively more likely. The search for causal mechanisms is indeed founded on this assumption.

In Chapter 4, I argue that wherever we are concerned with explaining the behavior of anything qua member of a natural kind, the search for causal mechanisms takes a particular form. It becomes a search aimed at discovering the real essence of the kind in question. What is the structure or composition of the kind, in virtue of which it is a member of this kind, and what causal powers does it have in virtue of its having this structure? Explanations that answer these questions I call *essentialist explanations*. They are essentialist because if they are true, they are necessarily true. A substance could not be a substance of the kind it is if it did not have this structure or these powers. The explanations of the chemical kinds are manifestly explanations of this sort. Insofar as these explanations are successful, they not only distinguish the chemical kinds from each other, they also explain the specific causal laws involving them.

At the most fundamental level of physical inquiry, it may be that the search for structure must drop out. For it may be that the most fundamental things have no structure, and therefore no structure in virtue of which they have the powers they have. In that case, we may have to distinguish between natural kinds solely on the basis of their causal powers. In other words, there may be causal powers that lack even a partial categorical basis. If this is so, then so be it. I see no objection other than David Armstrong's Meinongian one to this possibility (considered at the end of Chapter 3).

It is a corollary of this essentialist theory of explanation in the natural sciences that there can be no essentialist explanations constructed in any field where the subject matter is not naturally divided into kinds. Consequently, there can be no genuine causal laws in any of these areas. That is, there can be no proper causal laws of history or sociology or anthropology, a conclusion with which few would disagree. There is, however, a sophisticated body of theory – economic equilibrium theory – which bears most of the hallmarks of a genuine scientific theory, but is evidently concerned with social institutions that are not very plausibly members of natural kinds. So this branch of economic theory presents something of a test case for scientific essentialism. If the laws of nature derive from the essential properties of natural kinds, as the scientific essentialist maintains, then there ought not to be any genuine laws of economics. This challenge is taken up in Chapter 5, where it is argued that the so-called laws of economics have a status similar to that of the axioms of Euclidean geometry. They are true by definition or convention, and so have only a kind of *de dicto* necessity.

Chapter 6 begins a more focused discussion of the laws of nature. Laws are classified on the bases of scope and role. There are global laws which apply to all events and processes of the kind that can occur in our world, and the general structural principles that set limits to the kinds of structures that can exist in our kind of world. These laws are of the widest possible scope. There are also the structural laws, which pertain to specific kinds of substances, and the causal laws, which apply to specific kinds of events or processes. The essentialist theory of laws that has been developed to this point is one concerned with more specific causal laws and substantial structures.

It is argued that a satisfactory theory of the laws of nature would have to deal adequately with a number of important problems, which I call the Necessity, Idealization, and Ontological Problems. It should also lead to a solution to what I call the Demarcation Problem – how to distinguish between laws of nature and formal principles, such as the axioms of a geometry or an economic theory. Only the essentialist theory of laws, I argue, deals satisfactorily with all of these.

The most difficult problem for almost any theory of laws, apart from the conventionalist theory, is the Necessity Problem. What is natural necessity and how is it to be accounted for? For an essentialist, natural necessity is a species of necessity which is distinguished from other species, such as formal logical necessity and analyticity, by its grounding. Formal logical necessities hold independently of their subject-matter, and so are not grounded in that subject-matter. Such logical necessities remain true and necessary under all substitutions for their non-logical terms. Analyticities, on the other hand, are not grounded in any formal structures, and depend on the meanings of the terms involved. They are true, we say, in virtue of the meanings of words. Metaphysical or real necessities are not grounded either in the logical forms of statements, or in the meanings of words, but in reality. These necessities are of two kinds – those that are grounded in the individual real essences of things and those that are grounded in the natural kind essences of things. Those that are grounded in the individual real essences of things are such that the thing to which reference is made, whether by name or gesture, could not fail to have the property ascribed to it while still being the individual it is. Those that are grounded in the kind essences of things are properties which no individual could lack, yet still be a member of the kind to which reference has somehow been made.

According to some philosophers, the individual essence of a thing that is a member of a natural kind includes its kind essence. I think that this is probably too strong, but I remain undecided what weaker thesis to put in its place. I want to allow that a U^{238} atom could become a U^{239} atom without losing its identity as an individual (by absorbing a neutron), but I do not wish to allow that a horse could become a cow without loss of identity. My current view is that individual essences do not matter very much from the point of view of a scientific essentialist. For the only kind

11

of metaphysical necessity required for an explanation of the necessity of laws is that which is grounded in the kind essences of things.

The discussion of this and other issues concerning natural necessity will be found in Chapter 7, where it is argued that the essentialist theory of laws gives a far better account of natural necessity than any positivist, Humean, or neo-Humean one. The most pressing problems concerning the essentialist theory of natural necessity are (a) that metaphysical necessity, which like all of the other species of necessity implies truth in all possible worlds, seems too strong – surely the contrary of any law is always conceivable, and therefore possible, and (b) that while the account may be acceptable for specific causal and structural laws, it does not or cannot apply to the global laws. These issues are discussed and the problems resolved in a general theory of natural necessity.

The book concludes in Chapter 8 with an assessment of scientific essentialism as a general metaphysic. Just as Humeanism is a wide-ranging metaphysic with implications for almost every area of philosophy, so too is essentialism. It is not only a theory about laws of nature or about natural necessity. It is also about the nature of reality. It is a theory which is incompatible with Humeanism in a number of important ways, and its acceptance would have to involve some very considerable changes in how we think about the world, how we should model reality in the construction of our theories, what kinds of examples or counter-examples it is legitimate to use in philosophical reasoning, and how we should reason about things in science. It throws new light on the problem of induction, and it implies an important distinction between real and epistemic possibility which forces us to recognize that judgments of possibility based on ignorance do not have the same status as judgments of possibility based on our knowledge of the essential properties of natural kinds.

The most serious objection to scientific essentialism I have encountered comes from the theory of counter-factual conditionals. This objection will be discussed in Sections 8.7 and 8.8 in Chapter 8, following my development of an essentialist theory of conditionals. The objection is this: If truth conditions for such conditionals are to be based on real possibilities, rather than just epistemic possibilities, as surely they should be, then the truth conditions for some perfectly ordinary, and highly assertible, conditionals will turn out to be problematic. I have to admit the soundness of this objection. But then I never did think that one could provide adequate truth conditions for counter-factual conditionals, and I have always thought that "possible worlds" semantics for such conditionals and other modal claims were phoney. For such conditionals, and indeed for logic generally, I argued (Ellis, 1979) that what is needed is not a theory about their truth conditions, but a theory about their acceptability conditions, and that we should stop thinking of logic as the theory of truth preservation, and begin to think of it, as subjective probability theorists always have, as part of the theory of rationality.

Part One

Concepts

1

Concepts of Scientific Essentialism

The language we speak is a classificatory one: In computerese, it is digital rather than analog. We use analog techniques to convey information when it suits us – for example, we imitate people's gestures or tones of voice, and we draw pictures or display photographs of things when we find it too difficult or unhelpful to try to describe them. But mostly we communicate verbally, using words to refer to things and classificatory terms or descriptive phrases to characterize them.

Classifications are made of things in every different category of existence. We classify, and so characterize, objects, events, processes, properties, shapes, structures, substances, waves, thoughts, arguments, and so on. Indeed, we must do so if we are ever to talk about any of these things. Moreover, we recognize that there are similarities holding between ordered pairs, ordered triples, and so on, of things in each category, and sometimes even between things in different categories. We are thus able to discern relations of various sorts. We are also able to recognize similar and dissimilar similarities, and consequently make a number of second-order classifications of things, and relationships between things.

The classifications that we make are not arbitrary. It is true that people in different ages and in different cultures have sometimes classified things differently. But the basic classificatory systems of people, which are reflected in the languages we speak, are all similar – similar enough for simple descriptions of things in any one natural language to be translatable without much loss into any other. For this to be the case, there must be some common basis in nature for the classifications we make. Presumably, it implies the existence of a more or less common human nature (deter-

17

mining the similarities of our responses) and also the existence of some more or less objective similarities between the things that are generating these common responses. It would be extraordinary if the common classificatory systems of people arose somehow from diverse responses to objectively diverse kinds of things.

If the classifications that we make are not arbitrary, but are grounded in human nature and in the nature of the world to which we are responding, then how is this grounding to be explained? We reduce this to the question: How are objective similarities to be explained? The similarities of our classificatory systems would easily be explained if it could be argued successfully that we are similar beings responding to the same or to similar things in the world. But in virtue of what are things seen to be similar? There are two classical answers to the question: (1) that similar things have something in common, and (2) that there are objective relations of similarity on which our judgments of similarity are based. The first of these answers leads to what H.H. Price (1969) has called "The Philosophy of Universals," the second to what he calls "The Philosophy of Resemblances."

For reasons that will become apparent, I am a strong supporter of a Philosophy of Universals – I think that there are objective properties and relations that different things may have in common. But my theory of universals, which will be elaborated in Chapter 2, is not like any other theory of universals, at least not in all respects:

1. It is a "sparse theory" in the sense in which D.M. Armstrong's (1978, 1989) theory is. That is, it distinguishes between properties and relations, on the one hand, and predicates, on the other, and argues that there are many predicates, even many that are satisfied by things in the world (and are therefore true of them), which do not name properties or relations. I will also argue, although of course I cannot produce an example, that there are likely to be many properties and relations that cannot be ascribed to things using only the predicates that are now constructible in our language. That is, I suppose, as Armstrong does, that there is an ontology of properties and relations that is independent of language. I would also argue that it is one of the primary tasks of scientific investigation to discover and map this ontology.
2. It is an ontologically rich theory that recognizes the existence of several different categories of universals – *substantive* universals, whose instances are the members of natural kinds of objects or substances; *dynamic* universals, whose instances are the members of natural kinds of events or processes; and *property* universals, whose instances are tropes – the instantiations of real properties or relations.

3. It is a hierarchical theory in that it allows for the existence of hierarchies of universals of various degrees of generality, ranging from the most general category-wide universals (which the members of any given category of things must all instantiate) to the most specific universals within a given category whose instances must all be identical. The most specific species of universals in the property hierarchy are universals in the classical sense.

4. It is an Aristotelian theory rather than a Platonic theory. That is, it accepts that for universals, instantiation is a necessary condition for existence. This holds for all universals, whether classical or not. In other words, the commitment is to the Aristotelian thesis of *universalia in rebus* as opposed to the Platonic thesis of *universalia ante rem*.

5. Natural kinds clearly have a central place in this rich but sparse ontology of universals. For the various kinds of universals are all natural kinds of one sort of another. Substantive universals are natural kinds of objects. Dynamic universals are natural kinds of events or processes. Even property universals can reasonably be construed as natural kinds, the members of which are tropes or property instances. Quantities and other determinable property kinds may then be construed simply as generic universals whose infimic species are the determinates that fall under them.

1.2 NATURAL KINDS AND ESSENCES

Natural kinds are distinguished from other sorts of things in several ways.[1] First, they are distinguished by their *objectivity*. The distinctions between natural kinds are based on facts about their essential natures or structures, not on how we find it useful, convenient, or natural to classify them. Thus, membership of a natural kind is decided by nature, not by us; and the question of whether something is or is not a member of a given natural kind can never be settled just by fiat or arbitration. This question can only be settled by discovering whether what is to be classified has the essential properties or structure of the kind in question. It follows that the identity of a natural kind can never be dependent only on our interests, psychologies, perceptual apparatus, languages, practices, or choices. For if the identity of a kind depended on any of these things, then it might well be a kind of our own making, not one that exists in the world prior to our knowledge, perception, or description of it.

Second, natural kinds must be *categorically distinct* from each other. For they must be ontologically grounded *as kinds,* and exist *as kinds* independently of our conventions. Hence, where we are dealing with natural kinds, there cannot be any gradual merging of one kind into another, so that it becomes indeterminate to which kind a thing belongs. For if there were any such merging, we should have to draw a line somewhere if we wished to make a distinction. But if *we* have to draw a line anywhere, then it becomes *our* distinction, not nature's. Natural kinds must be *ontologically* distinguishable from each other.

Third, the distinctions between natural kinds must be based on *intrinsic* differences. That is, the members of two different natural kinds cannot differ only extrinsically, depending on how things in the world happen to be arranged or happen to be related to one another. If a thing's membership of a natural kind were to depend on its relationships to other things, then its membership of the kind would be an accidental matter. It would be a relationship which depended on its accidental circumstances. Therefore, if there are any natural kinds, they must exist as kinds independently of any such extrinsic relations, and their identities must be dependent only on the intrinsic natures of their members, not on what their extrinsic relations to other things happen to be. To illustrate: It might be the case that the only gold in the universe is to be found on earth. But the natural kind distinction between gold and other substances does not depend at all on this fact. A substance could obviously be gold, but not located on earth. For a substance to be gold, it must be constituted as gold. It must have those intrinsic properties that make it gold. Likewise, for a process to be meiosis, it must be constituted as meiosis, and involve the same kinds of substances changing in the same kinds of ways.

Fourth, if two members of a given natural kind differ intrinsically from each other, and these intrinsic differences are not ones that can be either acquired or lost by members of the kind, then they must be members of different species of the kind. This is what I call the *speciation requirement*. The isotopes of uranium, – U^{235} and U^{238} – differ intrinsically from each other. However, they both have the essential nuclear and electron structures of uranium and are therefore species of uranium. Electromagnetic radiation of frequency 2000 differs intrinsically from electromagnetic radiation of frequency 3000. However, electromagnetic radiation of either frequency is propagated according to Maxwell's equations, and both are species of electromagnetic radiation.

Fifth, if anything belongs to two different natural kinds, these natural kinds must both be species of some common genus. In other words, the memberships of two distinct natural kinds cannot overlap, so that each includes some, but not all, of the other, unless there is some broader genus that includes both kinds as species. The requirement is satisfied trivially if one of the two kinds is a species of the other. This is a feature of hierarchical structures generally, so I call this the *hierarchy requirement*.[2]

Sixth, natural kinds are distinguished from other sorts of things by their associations with *essential properties* and *real essences*. If what makes an object or process one of a certain kind depends only on its intrinsic nature, then any object or process that has this nature is necessarily one of this kind. Call this the *essentiality requirement*.

It is useful to distinguish between fixed and variable natural kinds. There are some natural kinds that have all of their intrinsic properties and structures essentially. This is true of the fundamental particles – for example, such things as these cannot acquire or lose intrinsic properties or structures without ceasing to be things of the kinds they are. I call these *fixed* natural kinds. There are other natural kinds that have some of their intrinsic properties or structures accidentally. A metallic crystal, for example, can become electrically charged, and so acquire a causal power it did not have before, without ceasing to be a metallic crystal of the kind it is. Metallic crystals are therefore members of *variable* natural kinds.

At the far end of the spectrum of variable natural kinds, there are the biological species. Plausibly, these may be regarded as clusters of closely related natural kinds, whose essences are their genetic constitutions (Wilkerson, 1995). But if they are clusters of natural kinds, they are a long way from being clusters of fixed natural kinds. For animals and plants are all highly variable in their causal powers and capacities, and can obviously gain or lose them, while remaining genetically the same.

The variability of a natural kind appears to be a function of its complexity. The members of the simplest natural kinds have zero variability. That is, they have all of their causal powers, capacities, and propensities essentially, and so cannot acquire or be given any new powers or lose any of the powers they have without ceasing to be things of the specific kinds that they are. One cannot, for example, teach a copper atom or a proton any new tricks. But things belonging to more complex kinds may have variable powers – powers they may acquire, lose, or have changed – without their ceasing to be things of the kinds they are. Crystals of various kinds are good examples. At a more complex level still, it appears that things may have *meta-causal powers*, – powers to change the powers they have, while remaining things of the kinds they are. Human beings, for example, while not all members of the same strict natural kind (since they have diverse genetic makeups), are certainly able to choose to do one thing rather than another, and thus change their behavioral dispositions.[3]

It follows from the essentiality requirement that membership of a natural kind cannot be due to superficial resemblance. If it has the essential properties and structures of the kind, then it must be a thing of that kind,

21

even if it looks different. Conversely, anything that is a member of a given natural kind must have the essential properties and structure of things of this kind. Therefore, the sets of properties or structures that define these natures and thus distinguish the natural kinds from each other, are distinctive of the natural kinds. Traditionally, such properties have been called the essential properties of the kind, and together they constitute what is known as the real essence of the kind. They are the properties or structures in virtue of which any object or process of that kind is the kind of object or process it is.

In what follows, it will be supposed that the objects, properties, and processes that exist most fundamentally in the world are all instances of natural kinds, and that these occur in natural hierarchies. I call this the "Basic Structural Hypothesis." (It will be explained more fully in Chapter 2.) Acceptance of this hypothesis has a number of significant implications. First, it commits us to the existence of a variety of kinds of real essences. If every natural kind has a real essence, as we suppose, there are real essences, not only of natural kinds of *objects,* as nearly everyone acknowledges, but also of natural kinds of *events and processes.* And if *properties and relations* can properly be regarded as natural kinds of tropes, which I think they can, then there must be essential properties of properties and relations too. Second, if natural kinds exist in hierarchies, as seems to be the case, then there must also be hierarchies of real essences. The more general kinds of objects, properties, relations, and processes must all have correspondingly general essential natures, and the more specific kinds, more specific natures. Third, the members of the various kinds must be distinguished from each other by their intrinsic properties and structures, not by their extrinsic relations. Consequently, if scientific research is concerned to discover the essential natures of things, it must focus on their intrinsic properties and structures. The special circumstances of their existence must be seen to be of only secondary importance. Fourth, any true statement attributing an essential property to a natural kind must in some sense be necessarily true. It must be, I shall argue, metaphysically, or as I shall say, really necessary.

An important, although not the only, aim of science is to discover what natural kinds of objects, properties, and processes there are, and to discover their real essences. For it is only when we know this that we can say what makes a thing the kind of thing it is. Hence, if we wish to understand our world, then we must discover what kinds of objects it contains, what kinds of properties they may have, in what kinds of ways they may be related, and in what kinds of ways they may act or interact. For when we have discovered all of these things, and we know what the essential natures of the various natural kinds are, then we should be able to go on to say what could or could not occur in any world made up of things of these kinds, acting, or interacting in the kinds of ways in which such things can, and can only, act or interact in our kind of world.

1.3 SPECIES, INSTANCES, AND TROPES

For every natural kind, there is an intentionally definable natural kind class, which is the class of its members. The set of all members of this class is the set of instances of the kind. The instances of any natural kind must be clearly distinguished from their infimic species. A species of a natural kind K is any natural kind whose membership is necessarily included in the class C(K) of members of K. An infimic species of a natural kind K is any species of K that has no sub-species. An infimic species of a natural kind is thus a universal in the same way that any natural kind is universal. It may have any number of members. An instance of a natural kind K, on the other hand, is a member of the class C(K), and it is not itself a natural kind, unless the natural kind of which it is an instance is a second-order, or higher-order one that has natural kinds of lower orders as instances.

The class C(K) of instances of K defines what I call the extension of K. From what has been said already, no two distinct natural kinds can have the same extension. The extension of a natural kind S may be included in that of another K, if S is a species of the kind K. The extension of S may also overlap with that of the kind K, provided that there is a more general natural kind of which both S and K are species.

It is important to distinguish between natural kinds that have their extensions in different categories of existents, something which is rarely done in the literature. Those I call *substantive* natural kinds have objects or substances as their instances, and the extension of any such natural kind must consist of objects or substances. Those I call *dynamic* natural kinds have events or processes as their instances, and the extension of any such natural kind must consist of events or processes. Those I call *property* kinds have the tropes of properties, relations, or structures as their instances, and the extensions of such kinds will be classes of similar tropes. We shall have more to say about each of these kinds of kinds as the argument progresses.

Of these different species of kinds, property kinds are likely to be the most controversial, since property kinds can exist as a distinct species only if their instances are, necessarily, neither objects nor events. What then are

23

property instances? They are not, I will argue, the infimic species of quantities or other generic property universals. Specific masses, charges, volumes, and the like are themselves universals and therefore unfit to be counted as instances of other universals, unless they be second-order ones. For reasons to be explained later (in Chapter 2), I reject the theory defended by Bigelow and Pargetter (1988, 1990) that quantities are really second-order universals. In my view, they are first-order generic universals, and the specific quantitative properties, such as specific masses, charges, volumes, and the like, are their infimic species. So what then are their instances? In my view, they are tropes.

A trope of a property, relation, or structure is an exemplification of that property, relation, or stucture in some particular object or event or group of objects or events. A trope of roundness, for example, might be the example of roundness that is to be found in a particular billiard ball, or in the moon, in the sun, or in any other round object. A trope of a specific color red (if we can pretend for a moment that redness is a natural property) is the instantiation of this color, which is to be found in some particular. It might be the display of redness that is in the skin of some apple or in that of some tomato. A trope of the relation of *having the same mass as* might be the example of this relationship, which may be found to hold between the masses of two particular protons or between two particular molecules. A trope of the structural property that all methane molecules have is the exemplication of the structure in any particular methane moleclule. If a given sample of gas contains a million methane molecules, then there are a million tropes of the methane structure present in this sample. If there are ten red tomatoes in a basket, all of exactly the same color red, there are ten tropes of this particular shade of redness present in the basket (or more, if one discriminates between objects more finely). If there are fifteen billiard balls on the table, then there are fifteen tropes of roundness instantiated in these objects.

The tropes of properties, relations, and structures are thus located *in* objects or events. Their identities as tropes depend on the properties, relations, and structures of which they are tropes, and also on the objects or events or ordered n-tuples of those things in which these properties, relations, and structures are instantiated. So a trope may be identified by an ordered pair <U, a> consisting of a property or relation universal U and an object or event or set of such things a in which the universal U is instantiated. This is not to say that a trope is such an ordered pair. Nothing as fundamental in ontology as a trope could possibly be an abstract entity like a set. A trope of any infimic species of any natural property has to be something like an elementary fact or state of affairs – for example, the fact of this property's being instantiated in this object or this relation's being instantiated by this ordered pair of things. It follows that the tropes of any property or relation U must always exist *in* things or sets of things – that is, in whatever things or sets of things have this property or stand in this relation, U. The tropes are never the things themselves that have these properties or relations. Nor are they the universals that are instantiated in them.

There is thus a distinction to be drawn between being instantiated in something and being instantiated by it. Property universals or property kinds are always instantiated in the things that have or display them. They are never instantiated by them. The object or event in which a trope is instantiated is the locus of this trope. But it is not the trope itself. The tropes are the property or relation instances that stand to the properties and relations of which they are instances in the same way as individual electrons stand to the substantive natural kind whose instances are electrons.

This distinction between being instantiated in something and instantiated by it will turn out to have important implications for ontology, as we shall see in the next chapter. In the case of property kinds, it is easy to confuse instances with infimic species, because any instance of a such a kind will also be an instance of an infimic species of that kind. Any instance of the color-kind *red*, for example,[4] such as that exemplified in a particular ripe tomato, will also be an instance of precisely this specific shade of red. But this specific shade of red is also a color-kind which may have many instances – for example, in other ripe tomatoes. So the shade of red is not an instance of the color kind *red*, but a species of it.

There are no such or similar problems with natural kinds that have objects or events as their instances. The substantive natural kind, *baryon*, for example, has at least two infimic species – the substantive natural kinds *proton* and *neutron*. Each of these infimic species has billions of instances, and all of these instances are instances of the generic natural kind baryon. There is no danger that anyone will confuse instances with species here. The instances are objects, the species are universals. The same with natural kinds of events. α- and β-decays are two species of radioactive decay processes. They are dynamic universals. The millions of such events that occur in nature, however, are not universals, and no one would be likely to think that they are. There is therefore a clear distinction to be made between species and instances in the cases of substantive and dynamic universals. Only the parallel distinction for the case of property universals is likely to prove difficult.

25

In what follows, I shall often refer to the infimic species of property universals as "classical," and whenever I speak of "tropes," it is the tropes of property universals that I have in mind. In fact, I shall not use the word "trope" in connection with substantive or dynamic universals, although I suppose there is no good reason, other than custom, why not. I shall simply refer to the instances of such universals as objects or events, as the case may be.

1.4 INTRINSIC PROPERTIES AND STRUCTURES

To discover the essential properties of any natural kind, we have to know where to look. One thing we know to start with is that the essential properties distinctive of any natural kind must be independent of the histories, locations, and surroundings of its members. The essential properties of any member of a natural kind must therefore be among its intrinsic properties or structures. Most of us have a rough idea of what is meant by an intrinsic property or structure. It is a property that something has independently of any other thing. But it is hard to say much more precisely than this what an intrinsic property or structure is. Several philosophers have tried to do so. But their attempts have so far not succeeded, or not succeeded in explicating a concept that is relevant to the theory or practice of science. The problem is that the concept of intrinsicality that has mainly been sought is a logical or acausal one – that is, the independence required for intrinsicality has been assumed to be *logical independence*. But this may not be an appropriate conception for the kind of dynamic world we live in, in which things are highly interdependent causally. What is needed for science is a concept of *causal independence*.

For most philosophers, shape is paradigmatically intrinsic. For the shape of any object is logically independent of anything external to it. However, the shape of an object at a given time is not necessarily the shape it has intrinsically, in the more interesting causal sense of this word. Let A be a stretched rubber band and G its manifest shape. Now some object might have this shape intrinsically. A steel band, for example, might have this shape independently of the external forces acting on it. But it is not the case that if any object has this shape, then it necessarily has it intrinsically. If the rubber band were causally isolated from its surroundings, it would not have this shape.

Intrinsicality in the causal sense is therefore not a property of properties, as many of those who have tried to explicate a logical concept of intrinsicality have supposed. This is because the very same property (having

26

the stretched rubber band shape) may be an intrinsic property of one thing, (for example, a steel model of a stretched rubber band), and not an intrinsic property of another (for example, the stretched rubber band itself). Causal intrinsicality must therefore be either a relation between a property and its bearer that is distinct from the normal bearing relation, or else a property of that relation. I suppose it to be a property of the bearing relation.

If an object has the property P intrinsically, then P must be a property that it does, in some sense, really have, even if what is manifested or displayed is something different. For we do not want to allow that an object might have the property P intrinsically even though it does not really have this property. If it has the property P intrinsically, then it must have it, even though it might not be what is actually displayed. For, in addition to having the property P, the object might be subject to the action of some distorting force F, so that what is actually displayed is not just P but the resultant of P and F. In the case of the rubber band, I say that the object has a certain shape intrinsically, a shape that is not in the circumstances actually displayed. What is actually displayed results from superimposing a certain distortion onto this underlying intrinsic shape.

The logical and causal concepts of intrinsicality are therefore very different from one another. The logical concept purports to distinguish between two essentially different kinds of properties – those that are necessarily intrinsic and those that are necessarily extrinsic. The causal concept, on the other hand, concerns the nature of the relation between a property and its bearer. On this conception, properties are not in themselves intrinsic or extrinsic; they are had or possessed intrinsically or extrinsically. To mark the distinction, let us reserve the term "internal" for properties that are intrinsic in the logical sense, and introduce the two-place predicate ". . . is intrinsically . . ." to denote this special "bearing" relation.

One philosopher who has tried to define internal properties is Jaegwon Kim. Kim's definition of an internal property is equivalent to the following:

D1: G is *internal* $=_{df}$ possibly some object x has G, although no contingent object, wholly distinct from x, exists.[5]

There are problems with this definition, especially if one is a nominalist and does not make a clear distinction between properties and predicates. For example, the property of being the sole object in the universe at t satisfies **D1,** but this property, if it is a property, is clearly not internal. But

27

even for realists about properties there are problems. Rotating objects certainly experience inertial forces, the effects of which can be measured. Would they experience such forces if nothing else existed? If Mach's Principle holds, and does so necessarily, inertial effects due to rotation should not be considered to be internal. But if Mach's Principle is contingent, or does not hold true, inertial effects due to rotation satisfy the condition **D1**, and must therefore be considered to be internal. This is highly counter-intuitive.

The internal/external distinction required for science is not adequately defined by **D1**. For what is needed is not a pre-theoretical logical distinction, but one which depends on our theoretical understanding of the world. This is so, because the intrinsic/extrinsic distinction employed in science is one of fundamental theoretical importance. It is not a distinction that can be made, as Kim has tried to make it, just by imagining the rest of the world away, because it is not a case of what holds just in our imaginations. It is a question of what is causally independent of what. Therefore the distinction has to be made in the light of the best currently available theories about what kinds of influences there might be on the things we have under investigation. And as our theories change, so might the intrinsic/extrinsic distinctions we make change along with them.

Consider mass. It is now generally believed that the mass (rest mass) of an object is an intrinsic property of that object. That is, it is a property that the object has intrinsically and that it would display if it were viewed from the standpoint of its own frame of reference. It is not an *a priori* truth, however, that this should be so. If Mach had been right, the dynamical properties of bodies, including their rest masses, would depend on the distribution of matter in the universe and on their positions in relationship to it. In that case, rest mass would be an extrinsic property.

To capture the causal concept of intrinsicality that is needed, we might start by considering what properties an object would display in the absence of any external influences (such as those due to force or motion). Thus, we might attempt to define causal intrinsicality:

> **D2:** G is a *causally intrinsic* property of an object $x =_{df} x$ would display G, if there were no external influences affecting the properties displayed by x.

According to this definition, the intrinsic shape of a solid object would be the shape it would have in the absence of any external distorting forces.[6] Hence, given this definition, an object might have some particular intrinsic shape, even though it does not, and never will, actually display this shape. For example, the object might always be distorted by gravitational or other forces.

The definition **D2** comes close to capturing the causal concept of intrinsicality that is needed for science. Nevertheless, it is still not quite right. A solid object that is vibrating or rotating will also be distorted. But it will be distorted not by external forces, but by accidental ones. Therefore, for a property to be intrinsic, it must not only satisfy the definition **D2**, it must also be a property that a body would have if it were not subject to any such *accidental* forces – forces that may or may not be present on a given occasion. Since external forces are also presumably accidental, we can strengthen **D2** in the required way by substituting "accidental" for "external." Accordingly, we accept the following definition:

> **D3:** G is an *intrinsic* property of an object $x =_{df} x$ would display G in the absence of any accidental forces that might otherwise affect the properties that would be displayed by x.

It follows from this definition that unless a body happens to be at rest in an inertial system in deep space, and neither rotating nor vibrating, it cannot continually display the shape that is intrinsic to it.

This concept of intrinsicality is needed for science because it reflects an important part of the structure of causal explanations. To explain the shape of the earth, for example, we must contrast its actual shape with the shape it would have if it were not distorted by tidal forces, not rotating on its axis, and not vibrating. This is the earth's intrinsic shape. Its intrinsic shape is not, of course, spherical. For the earth to be intrinsically spherical, it would have to be perfectly homogeneous. But we know that that is not so.

Roughly, the distinction we need for science is between what is due to internal static forces alone and what is due, wholly or partly, to external forces or circumstances. As we have seen, this is not the whole story, because the internal properties of things can be distorted by other processes. However, for many scientific purposes, the distinction between what is due to internal forces alone, and what is not, is sufficient. This distinction, in these terms, is fundamental in the physical sciences, for example, because it is needed to state the conservation laws. These laws all have the following form: Every event or process which can occur in nature is intrinsically X conservative (where "X" names the conserved quantity). The distinction is also needed for the construction of certain kinds of physical theories. The abstract theoretical models of physical systems are often designed precisely to represent the properties and structures of systems in causal isolation from their surroundings.

It is not true to say that all theoretical models of physical systems are designed to represent the properties and structures that physical systems would have if they existed in causal isolation. It depends on what is to be explained by them. If what is to be explained is a system of relations that, let us say, does not depend at all on the intrinsic natures of the things related, then the appropriate theoretical model for this system of relations will reflect this fact. Our theories of space and space-time, for example, have nothing to do with the intrinsic natures of the things that may be spatially or spatio-temporally related. Consequently, these things may be represented in our abstract theories by points or point-events – theoretical entities that have no intrinsic properties or structure, only location.

Other theoretical models are concerned with the patterns of interaction between two or more physical systems, and hence not only with the intrinsic properties of these systems considered in isolation. Nevertheless, these interacting systems may constitute a complex system whose elements are intrinsically disposed to interact in certain ways. A theoretical model of any such complex system will normally abstract from the various external forces that may act on the whole complex, or on any of its parts, to focus on the internal forces acting between the parts. Any external forces that may be acting on the system would be merely disturbing influences that are strictly irrelevant to the theory of the interaction. Consequently, they would quite properly be ignored in the theoretical model that is required to explain this interaction.

The properties that are of most concern in the physical sciences are those properties a thing has independently of the contingencies of its location, and its relations with other things, and would display if it existed as a closed and isolated system. For these are the properties it has intrinsically. They are not necessarily the properties it would actually display in normal circumstances, since the properties normally exhibited by things are generally due to a mixture of causes, both internal and external. The properties that a thing has intrinsically are just the properties that it would display if the internal forces were the only relevant factors. Some will raise the objection that to distinguish between intrinsic and extrinsic properties in this way leaves us in a circle of inter-definable properties.[7] For the distinction I now want to make depends on that between internal forces or causes on the one hand and external causal influences and relations on the other. How is this distinction to be made?

The intrinsic properties and structures of things are what make them what they are. They explain how things are disposed to behave, just in virtue of how they are constituted, and what the causal powers of their constituents are. But this is precisely what Lockean real essences are supposed to do. Therefore, the intrinsic properties and structures of things are their Lockean real essences.

It is important to understand that things could have Lockean real essences even if no natural kinds existed. If the world were, as Locke imagined it to be, a mechanical world without quantum phenomena, things might well have Lockean real essences – properties and structures that make them what they are, even though any one kind of thing might be transformed by gradual degrees into any other. The rainbow colors, for example, have Lockean real essences. For we can say clearly what makes any given sample of monochromatic electromagnetic radiation in the visible part of the spectrum the spectral color it is. But the spectral colors are, nevertheless, not natural kinds, because they violate the distinctness requirement.

Therefore, for natural kinds to exist, it is not enough that things should have Lockean real essences. Things belonging to different natural kinds must have clearly distinct real essences.[8] There cannot be any borderline cases between the real essences of different natural kinds because, if there were, the distinctions between the kinds would be superficial, as, for example, the blue/green distinction is, having no basis in the underlying reality. For natural kinds to exist, there must be discreteness or discontinuity at the most fundamental level. If objectively different kinds of things exist in nature, the distinctions between them must be there in their intrinsic natures for us to discover.

Many of the natural kinds that are of most interest to us are ones whose members are intrinsically identical to each other vis à vis their standing properties, but intrinsically distinct from all other kinds of things in these respects. For these are the natural kinds that appear to be ontologically most fundamental. For example, any one electron is intrinsically identical to any other electron in respect of its standing properties, but is intrinsically different from every other kind of particle in some or all of these respects. Let us say that a natural kind is a *simple natural kind* iff its members must all be intrinsically identical in this way.

The class of simple natural kinds includes the neutrons, the protons, the neutrinos, and all of the other distinct kinds of fundamental particles. It also includes some more complex structures. The various atomic nuclei, the neutral atoms of all of the isotopes of all of the elements, the molecules of each of the chemical compounds formed from the same isotopes, and so on are all simple natural kinds, as far as we know. For the members of these classes would appear to be intrinsically identical vis à vis all of their standing properties. It is true that atoms and molecules can exist in various states of excitation. But excitation states are not standing properties; they are properties that a thing can acquire or lose without ceasing to be a thing of the kind it is. They are what essentialists call "accidental properties." The class of simple natural kinds is thus a very broad class which includes most of the kinds of things we believe to exist most fundamentally. Indeed, it is plausible to suppose that the substance of the world consists entirely of things of such kinds.

Most kinds of things that we can reasonably suppose to be natural kinds are not simple, however. Some are generic kinds, which have species. The copper atoms, for example, include the various isotopes of copper as species. Some classes of things that are plausibly natural kinds are really clusters of similar natural kinds, which are conceptualized as being things of the same kind, although there is no set of standing intrinsic properties or structures that would distinguish the members of these kinds from those of all other kinds. These are what we might call 'cluster kinds'. The members of the various animal and plant species, for example, are members of cluster kinds.

1.6 TWO BRANDS OF NECESSITY

In the philosophical literature, real essences are traditionally distinguished from nominal essences. The nominal essence of a kind (not necessarily a natural kind) is said to be the set of properties in virtue of which a thing is called a thing of that kind. It thus relates to the process of classifying things. The nominal essence of a kind defines, or is part of the definition of, the word that designates the kind. The real essence, by contrast, is said to exist independently of language, and to be discoverable by empirical investigation. The real essence is sometimes said to define the kind itself, not just the word used to designate it.

The distinction between real and nominal essences is usually associated by philosophers with another distinction, no less important, between two brands of necessity. Classically, it is the distinction between *de re* and *de dicto* necessity. More commonly, the distinction is said to be between what is metaphysically or really necessary on the one hand and what is analytic, or true by definition or convention, on the other. This distinction is one that philosophers will no doubt argue about. But the need for some such distinction is widely recognized. For the laws of nature would appear to be both discoverable by scientific investigation, and, at least in some sense, necessary. But if they are necessary, then in the absence of any such distinction, it seems that we should have to say that the laws of nature are analytic, or true by definition or convention. And, this would seem to imply that the laws of nature are not discovered, but stipulated!

It will be argued presently that propositions that are necessary in either of these two senses are true in all possible worlds. For, strictly speaking, the distinction between the two brands of necessity is one of grounds, rather than modality. *De dicto* necessities are grounded in the conventions of language; *de re* necessities are grounded in the natures of things. Nevertheless, the *de re/de dicto* distinction is an important one that must be kept clearly in mind if the thesis presented in this book is to be properly understood. I will be arguing, for example, that the laws of nature are all metaphysically necessary, and therefore both necessary in the full-blooded sense of this term and grounded in nature. They are not contingent, as most philosophers believe. Nor are they just conventions adopted for the purpose of organizing our empirical knowledge of the world. Nor are they necessary in any weak sense of necessity. The laws of nature are metaphysically necessary, I shall argue, because they are all true in virtue of the essential properties of the natural kinds of things existing in the world.

Propositions that are not metaphysically necessary, but necessary for any other sorts of reasons, will all be classified together as necessary *de dicto,* although I do not claim that all propositions that are necessary *de dicto* are necessary or true for just the same sorts of reasons. In the simplest kind of case, a proposition is necessary *de dicto* because it is straightforwardly true by verbal definition. The proposition that a bachelor is an unmarried man, for example, it necessarily true for this reason. It is not necessarily true because there is a natural kind class of bachelors whose members are unmarried men. Bachelorhood fails all of the tests of natural kind-hood. It is true because the term "bachelor" is used to pick out a certain class of men in which we happen to have some interest. The proposition that electrons are negatively charged, by contrast, is not true for any such superficial reason. Electrons exist as members of a natural kind, whether or not we have any knowledge of or interest in them. And these things, which we call "electrons," are intrinsically identical objects whose distinctive intrinsic properties include negative charge. The proposition is thus one that

33

is true in virtue of the essential natures of the things in this class, and as such it is a proposition that would be true even in a world without people. For nothing could be an electron if it were not negatively charged, and this would be the case even if there were no language in which this fact could be stated.

Propositions such as "A bachelor is an unmarried man," which can be shown to be true just by an analysis of the meanings of words, are said to be *analytic*. As such, they are contrasted with *synthetic* propositions, which are not derivable from an analysis of meanings. But not all propositions that are true by definition or convention are analytic in this straightforward way. Some, such as the proposition that the speed of light is the same in all directions, are supposed to depend partly on facts and partly on definitions. Others, such a "Red is a color," are problematic, because terms like "red" and "colored" are defined only ostensively, and it is hard to see how anything much can be deduced from ostensive definitions. On the other hand, there are some propositions that are evidently true in virtue of their logical form. And these propositions are also widely regarded as being true by definition or convention. Of these propositions, it is said that they are true in virtue of the meanings of the logical connectives and operators. It is not obvious, however, that this makes them true or necessary in the way that the standard examples of analytic propositions are true or necessary.

I do not take sides on any these issues. For the question that concerns me here is not whether there is a simply definable class of *de dicto* necessary propositions, but whether there is a recognizably distinct class of metaphysically necessary ones.

What then of the proposition that electrons are negatively charged? Isn't this just an analytic proposition? No, there are important differences between this proposition and the proposition about bachelors that I have taken as a paradigm of analyticity. First, if someone does not know that electrons are negatively charged, it does not follow that he does not know what electrons are. He might know, for example, that they are the lightest of the fundamental particles in the family of leptons, that they are emitted in the process of β-decay, that they are the same as cathode rays, that they have spin (intrinsic angular momentum) 1/2, unit mass, and so on. He might even know that they are the orbital constituents of atoms, that ionization involves the loss of electrons, that their anti-particles are positrons, and many other things besides. Of course, it is extremely unlikely that if he knew all this, then he would not know that electrons are negatively charged, since this is such a basic property of electrons. But it is possible.

However, the case is very different with bachelors. If someone does not know that a bachelor is an unmarried man, then he does not know what a bachelor is. For bachelors have no characteristics, other than their defining characteristics, that would distinguish them from other men. There could perhaps be generalizations that are true of all and only bachelors, although it is hard to think of any convincing examples. For the sake of argument, let us allow that bachelors are generally gregarious people, while married men are generally solitary. If this were true, it would no doubt be a generalization about bachelors of some sociological interest. But a person who thought that bachelors were just gregarious men, whatever their marital status, would not know what a bachelor is. To know what bachelors are, one needs to know that they are unmarried men. Nothing else will suffice.

But is the case of the electron really so different from that of the bachelor? It is true that the concept of an electron is a complex one, and that there are many different properties an object must have to be an electron. But does this show the case to be so very different? Let $P_1, P_2, P_3, \ldots P_n$ be the essential properties of electrons. Should I not then simply define an electron to be any object that has all of the properties $P_1, P_2, P_3, \ldots P_n$? And if P_1 is the property of being negatively charged, would it not then be just an analytic proposition that electrons are negatively charged? My answer is no. First, such a *real definition,* as it might be called, is corrigible in a way that a *nominal definition,* such as the definition of a bachelor as an unmarried man, never is. Empirical findings could show that something that we now take to be an essential property of electrons, and therefore part of its real definition, is not, after all, such a property. If, for example, electrons were found to differ from each other very slightly in mass, just as the atoms of the different isotopes of chlorine do, then we should have to allow that electrons may differ essentially from each other in certain ways. We might even have to concede that nothing has the specific mass of the electron (just as chemists had to concede that nothing really has atomic weight 35.5). But the definition of "bachelors" is not thus vulnerable to empirical refutation. Nothing that might be discovered about bachelors could show that bachelors are not, after all, unmarried, or men. Second, real definitions, unlike nominal definitions, are open-ended. The known essential properties of electrons might be $P_1, P_2, P_3, \ldots P_n$, but there may well be other essential properties that are yet to be discovered, and these would also have to be included in the real definition. There is a sense, therefore, in which it is probably true to say that we do not yet know what electrons really are.

But the main reason for distinguishing between a metaphysically necessary proposition and an ordinary analytic one is that the ground of the former is radically non-linguistic and objective, while that of the latter is not. Neither the truth nor the necessity of the proposition that electrons are negatively charged depends on either the language or the distinctions customarily made by speakers of the language. Even if we don't have a name for electrons, and electrons are not distinguished in the culture from other kinds of things, it is still the case that electrons are negatively charged. If, for example, we could not refer to electrons, other than ostensively as "that kind of particle," they would still necessarily be negatively charged. For that is something that is essential to this kind of particle. We might say, for example, that thingummies – we do not know what they are called, or even whether they have a common name – are negatively charged. And this proposition, if it is true, and if "thingummies" refers ostensively to electrons, is necessarily true. But obviously the proposition that thingummies are negatively charged is not analytic.

By contrast, the proposition that a bachelor is an unmarried man is linguistically dependent, or at least is dependent on our social practices. For the class of unmarried men cannot be ostensively defined and cannot be named as other than the class of bachelors or the class of unmarried men. Therefore, the proposition that thingummies are unmarried men is not, and cannot be construed as, an analytic proposition.

I have chosen to argue the case for the existence of a distinct brand of necessity – namely, metaphysical necessity, by choosing, as an example of a metaphysically necessary proposition, one that is as much like an analytic proposition as possible. I could have chosen an example that is more obviously not analytic but no less evidently metaphysically necessary. Consider, for example, the proposition that an electron has spin equal to 1/2. This proposition is certainly metaphysically necessary because the property of having spin 1/2 is an essential property of electrons. But it is clearly not analytic.

The proposition that an electron is a particle with spin 1/2 is one of considerable theoretical importance about electrons, and it cannot be derived just from an analysis of the meanings of words. If we found something that was otherwise like an electron, but had spin other than 1/2, we should have to allow that there are species of electron-like particles that differ from each other in spin. And this would pose a serious theoretical problem: Why should two particles, otherwise intrinsically identical as far as we can tell, differ in only in this respect? What are these new particles with anomalous spin? What would be their place in the family of fundamental particles?

But such theoretical issues cannot arise with bachelors. For the distinction between bachelors and other men depends on relations between people, and these relations are, by nature, accidental relations. That is, they are relations that might or might not hold of a given man at a given time. Intrinsically, bachelors and other men are much the same. At least, they are intrinsically no more different from each other than bachelors and other men are among themselves. But contrast this with the property of having spin 1/2. This property is not just an accidental property of a fundamental particle; it is not a property that a fundamental particle of a given kind might or might not have, depending on the circumstances. It is not a variable or manipulable quantity, but a property that a particle has intrinsically, in virtue of its nature. There is no known way of changing the spin of a particle; it is just as invariant as its charge or mass. Whatever its value, then, the spin of a particle would appear to be one of its essential characteristics.

The difference between a metaphysically necessary proposition such as "an electron has spin 1/2" and an analytic proposition such as "a bachelor is an unmarried man" is not so much a difference in the kind of necessity as a difference in the kind of grounding of the necessity. Both propositions are, as logicians would say, true in all possible worlds. But while the metaphysical necessity is grounded in the essential properties of a natural kind, the *de dicto* necessity of the analytic proposition is grounded in the definition of a socially constructed class. The necessity of the metaphysically necessary proposition thus depends on what exists in nature, whereas the necessity of the analytic proposition depends on our social practices and linguistic conventions.

The distinction between metaphysical necessity and truth by definition or convention is nevertheless a very important one. For, according to the essentialist theory I wish to defend, the laws of nature are metaphysically necessary. They are not true by definition or convention, and so necessary *de dicto,* as the conventionalists maintained. They are not just true empirical generalizations, and so not really necessary at all, as the Humeans believe; nor is their necessity due to any mysterious relations of natural necessitation between universals, as several modern theorists have argued.

37

The necessity of laws is grounded in the essential properties of natural kinds of objects, properties, relations, and processes. The laws of nature describe the essential natures of such things.

1.7 NECESSITY AND *A PRIORI* KNOWLEDGE

Philosophers have traditionally distinguished between *a priori* and *a posteriori* knowledge. What is *a priori* is what can be known before any empirical investigation begins, and so is logically prior to the results of our investigations. What is *a posteriori* is the opposite. It is what can only be known by experience, or as a result of experience.

The traditional problem of *a priori* knowledge is that which was thought to be posed by geometry and arithmetic. For geometrical and arithmetical knowledge has appeared to many philosophers of earlier generations to be both *a priori* and synthetic. Kant's *Critique of Pure Reason,* for example, was written specifically to answer the question: How is such synthetic *a priori* knowledge possible? There is no difficulty in understanding how analytic propositions can be known *a priori*. For analytic propositions are by definition propositions that can be known to be true just by analyzing the meanings of the words used to express them. There is therefore no need to investigate anything to which the terms in the proposition might refer to discover whether an analytic proposition is true. Indeed, if one takes a proposition such as "a bachelor is an unmarried man," it is hard to see how a relevant investigation could even begin. If one does not already know that a bachelor is an unmarried man, then one cannot identify the people to be investigated. So one cannot begin to investigate the properties of bachelors.

Kant's solution to the problem of synthetic *a priori* knowledge was to argue that our perception and understanding of the world are conditioned by a certain in-built system of concepts and principles whose natures can be discovered just by reflection on how we are able to reason and think about the world. Experience is not needed to validate the knowledge to which these concepts and principles give rise, because our possession of them is a necessary condition for the possibility of any experience whatsoever. Kant concedes that all our knowledge begins with experience, but argues that it does not all arise out of experience.

The details of Kant's solution need not concern us here. However, there is one Kantian claim which came to be very widely accepted by empiricists and indeed by philosophers of science generally. Kant argued that there are two marks of the *a priori* – necessity and strict universality. I cannot do better than quote Kant himself on this:

What we here require is a criterion by which to distinguish with certainty between pure and empirical knowledge. Experience teaches us that a thing is so and so, but not that it cannot be otherwise. First, then, if we have a proposition which in being thought is thought as *necessary*, it is an *a priori* judgement; and if, besides, it is not derived from any proposition except one which also has the validity of a necessary judgement, it is an absolutely *a priori* judgement. Secondly, experience never confers on its judgements true or strict, but only assumed or comparative, *universality*, through induction. We can properly only say, therefore, that, so far as we have hitherto observed, there is no exception to this or that rule. If, then, a judgement is thought with strict universality, that is, in such a manner that no exception is allowed as possible, it is not derived from experience, but is valid absolutely *a priori*. (Kant, 1787, 43–4).

This strong linking of necessity and strict universality with *a priority* has had a profound influence on the history of philosophy. Once it is accepted, it creates a number of major philosophical problems. First, there is Kant's problem of the existence of synthetic *a priori* knowledge. Second, and most important from my point of view, there is the problem of the status of laws of nature. If the laws of nature are necessary and strictly universal, they must be *a priori*, or at least derivable from *a priori* principles. If they are not necessary or strictly universal, then what is their status, and why should we believe in them? Are they, perhaps, just universal generalizations to which, as yet, no exceptions have been observed?

Kant's linking of necessity with *a priority* had been anticipated by Hume. For Hume had argued for the contrapositive of Kant's thesis – that whatever is knowable only *a posteriori* must be contingent, and so is not necessary. No amount of experience, he said, however much or consistent it may be, can ever show that one event is necessarily connected to another. To show that there is a necessary connection between two events, we should need an argument that the one must follow the other. But, for distinct events, no such arguments are to be found. The events are not logically connected, so any argument for necessary connection would have to be based on experience. However, there are no arguments from experience that can show that there is a necessary connection between any two events.

The situation would perhaps be different if two events could be shown to be causally connected. For then we could argue from the principle that a cause must be sufficient to produce its effect to the existence of a necessary connection. However, there are no legitimate concepts of causation, force, or power applicable to events from which the idea of necessary connection could arise. For we have no impressions of any such things in nature. We can observe that one thing is regularly followed by another, perhaps, and is always, in our experience, conjoined with it, but we can never observe any power or force connecting them.

Hume's philosophical methodology was always to inquire what the sources of any supposed idea might be. If it is a legitimate idea, he argued, then it must have its origins in our impressions (including what Hume calls our 'inward sentiments'). Hume reasoned, therefore, that our ideas of causal or necessary connections between distinct events can never arise from any objects or sequences of events that we can ever observe. Connections of these kinds are simply not to be found in nature. Consequently, he argued, to make sense of the idea of one event causing, or necessitating, another, the required concepts of causation and necessitation must be grounded somehow in us, rather than in the things we are observing. He supposed them to be grounded in the habits of thought which are induced in us by our repeated observations of regularities in nature, and in the expectations to which they give rise. For he could find no other possible source of these ideas. Hume thus came to the conclusion that the laws of nature are nothing more than contingent general propositions about the world, and if they appear to us to be necessary, then this is just because we are projecting our own subjective expectations or convictions on to the world.

Clearly, if Hume and Kant are right that what is necessary must be *a priori* and that what is *a posteriori* must be contingent, then this poses a dilemma concerning the status of the laws of nature. If the laws of nature are *a posteriori,* then they are contingent; if they are necessary, they are not *a posteriori.*

Naturally, philosophers have taken different sides on this. According to one group of philosophers, the laws of nature are true, but falsifiable, empirical generalizations about the world. They are contingent, *a posteriori,* and true universal propositions. We call this theory of laws of nature "the Humean theory," for this is roughly what Hume believed. The theory was elaborated and defended by K.R. Popper, and has been strongly supported more recently by J.J.C. Smart and other scientific realists.

The contrary view that the laws of nature are *a priori* and necessary has not to my knowledge been defended by anyone in recent years. It is now just too implausible to claim that the laws of nature are *a priori*. However, it was widely believed in the seventeenth and eighteenth centuries that it was possible to have a pure science of mechanics, just as one could have a pure science of geometry or arithmetic. Indeed, Kant himself evidently believed that a pure science of mechanics was possible. By a pure science, Kant meant a body of scientific knowledge that was logically based, as geometry was then thought to be, entirely on *a priori* intuitions. And since at that time the laws of nature were closely identified with the laws of mechanics, we may reasonably call the theory that the laws of nature are both *a priori* and necessary "the Kantian theory."

The Humean theory has survived almost in its original form, although very few philosophers these days would accept Hume's psychologistic explanation of the apparent necessity of laws. On Hume's theory, the following two events are "loose and separate": A – Jemima the cat's being flattened by a falling piano, and B – the death of Jemima. According to Hume, the idea that A is the cause of B, or somehow *brings B about* is really based on the association of these two events in our minds. The events themselves are not necessarily connected. It is just that the thought of the one, A, inevitably *suggests* the other, B. This is claimed to be so, because whenever in the past we have observed a heavy object, such as a piano, fall on and flatten a cat, such as Jemima, the cat has died. We are thus habituated to expect Jemima to die when she has been flattened by a piano. One would indeed have to be a very devout follower of Hume to accept this explanation as adequate.

While the Kantian theory of laws as *a priori* truths has not survived, the related theory that some, if not all, of the most fundamental laws of nature are necessary because they are true by definition or convention became very widely accepted. Indeed, this kind of conventionalism was the dominant theory of the laws of nature for most of the last century, and has only seriously been challenged in the last twenty years or thirty years.

Conventionalism rested on a fundamental distinction between pure and applied science, just like the distinction between pure and applied mathematics. The laws of nature were seen as belonging to pure science, and consequently to have roles and a status similar to those of the axioms and definitions of a geometrical or other mathematical system. Within the system, they are inviolable. No experience could ever refute them. The only question is whether the system is the best, the simplest, or the most appropriate for describing the world.

Of course, the purposes of theory construction in science are different from those of theory construction in mathematics. For a scientific theory, unlike a mathematical theory, must have an intended application by which its worth or appropriateness can be judged. So the laws of nature, which are identified as the axioms or definitions of global scientific theories, are constrained by the requirement of applicability. They are conventional, certainly. But they are not arbitrary conventions. They are chosen specifically for the purpose of organizing and systematizing our experience of reality in the best, and simplest way. Here is Poincaré on the subject of some basic laws of classical mechanics:

Are the laws of acceleration and of the composition of forces only arbitrary conventions? Conventions, yes; arbitrary, no – they would be so if we lost sight of the experiments which led the founders of science to adopt them, and which, imperfect as they were, were sufficient to justify their adoption. It is well from time to time to let our attention dwell on the experimental origin of these conventions. (Poincaré, p.110).

Most conventionalists did not say that all laws are conventions. For clearly there are generalizations that are empirically discovered and empirically refutable. But these were never held to be necessary truths, and therefore they did not require conventional status to explain their necessity. Conventional status was attributed only to the most fundamental laws of nature – those that were clearly embedded in a body of theory, and integral to it. The empirical findings of botanists or chemists were not thought to be laws of nature in this sense, but only the general facts that are to be explained by any adequate botanical or chemical theories.

In this book, I argue that neither a Humean nor a conventionalist theory of the nature of laws is satisfactory. Humean theories fail because they cannot adequately account for the necessity of laws. Conventionalist theories fail because they are unable to account for the success of theories. If there is no underlying structure to ground our successful theories, why do they work? These points will be argued in Part IV. The basic reason for the failure of both theories, it will be argued, is that they fail to distinguish between real and epistemic modalities – between real necessities and possibilities and epistemic necessities and possibilities. Real possibilities exist in nature and have to be discovered. What is epistemically possible, however – possible for all we know – might not really be possible, and what is really necessary might not be epistemically necessary. Once this distinction is granted, it becomes clear that the supposed necessary connection between *a posteriority* and contingency, which was accepted by both Hume and Kant, must be rejected. For it is not the case that whatever can be known only by experience, and is therefore epistemically contingent, cannot also be really necessary.

There is an important difference between generalizations that just happen to be true and laws of nature that in some sense, tell us what must be the case. Generalizations such as "all of the books on Caroline's desk are philosophy books" do not imply that the only books that could be on her desk are philosophy books. But laws of nature such as: "To every action there is an equal and opposite reaction" do imply that the only actions that could occur are ones that are accompanied by equal and opposite reactions.

To mark the difference between these two kinds of generalizations, philosophers distinguish between *accidental* and *nomic* universals. The generalization that all of the books on Caroline's desk are philosophy books is an accidental universal, whereas the principle of action and reaction is a nomic universal. The distinction is usually made in this way: Nomic, or lawlike, universals support subjunctive conditionals; accidental universals do not. Thus, from the proposition that all of the books on Caroline's desk are philosophy books, it does not follow that if the Oxford English Dictionary had been on Caroline's desk, then it would have been a philosophy book. But from the principle of action and reaction it does follow that if the bullet had been fired from this gun, there would have been a reaction equal and opposite to this action.

The kind of necessity that nomic universals evidently have is called *nomic* or *natural* necessity. It is one of the main problems in the philosophy of science to describe this kind of necessity and explain how it arises. Humeans obviously have difficulty with it. If the laws of nature are supposed to be nothing more than large-scale generalizations, no different in kind or principle from the less-comprehensive generalizations that everyone agrees are accidental, the Humean is committed to saying that their necessity does not lie in the things over which these generalizations range. Therefore, if the necessity exists at all, it must arise from our manner of thinking or speaking about these things.

Conventionalists also have trouble with natural necessity. For, like the Humeans, they cannot allow that it has any real basis in nature. For they do not think there are any real (that is, *de re* or metaphysical) necessities in the world. All necessities, they suppose, are *de dicto,* and hence derive from the conventions of language.

Nature makes some ways of speaking or thinking about the world better, or more convenient, than others. Otherwise, we could never agree about anything. But conventionalists cannot explain what makes some theoretical structures so much better than others that could, in principle, be constructed. For example, we could, as Reichenbach has argued, describe the world as a Euclidean space of three dimensions, with a fourth dimension of time, provided that we were prepared to make radical enough changes elsewhere in our physics to fit in with this classical assumption. However, the resulting system would be so monstrously complex that no one would give it a moment's consideration.

Why should this be so? A realist about curved space-time can reply that the classical assumption that space and time are Euclidean is simply false. Therefore, if we make this false assumption, we shall have to make a number of other false assumptions to account for the observed facts. So, like a liar in the witness box, we shall have to fabricate falsity after falsity to maintain the coherence of our story. But this reply is not available to a conventionalist. For conventionalists are committed to the view that laws and theories are neither true nor false. There is no truth of the matter, Reichenbach said, whether space is Euclidean or not. It is just a matter of what theoretical system we find most convenient.

The best way to explain the appearance of natural necessities, I argue, is to postulate their real existence. Of course, I am not the first to have argued this. Others who have wrestled with the problem of natural necessity have come to a similar conclusion. But, for the most part, those who have done so have attempted, disastrously, to remain within the Humean framework. They have accepted what Caroline Lierse and I call the Humean metaphysic.

1.9 THE HUMEAN METAPHYSIC

In Hume's ontology, there is a sharp distinction between what kind of thing something is and how it is disposed to behave. The kind of thing it is is supposed to depend only on those of its properties that are both intrinsic and categorical – intrinsic in the sense that they do not depend on relations with other things, and categorical in the sense that they do not depend on what the laws of nature are. Thus, it is assumed that things belong to different kinds or categories, independently of their relations to other things, and independently of how they are disposed to behave. On the other hand, how a thing of a given kind is disposed to behave is supposedly dependent on the laws of nature. If the laws of nature concerning things of this kind were different, then these things would be disposed to behave differently.

Thus, on Hume's theory, there are no intrinsic dispositional properties – properties in virtue of which things must, in appropriate circumstances, behave in some specific way. The behavior and actions of things are supposed to be determined, or made probable to some degree, not by their intrinsic properties (all of which are supposed to be categorical), but by the laws of nature as they apply to things having these properties.

Dispositional properties, which dispose things to behave in certain ways, either by necessity, or with some degree of probability, are sometimes called *modal* properties. For they are concerned with what things must, could, or could not do in various kinds of circumstances. Therefore, we may make the point of the previous paragraph by saying that in the Humean metaphysic, there are no modal properties that do not depend on categorical non-modal properties. All modal properties, it is said, must have categorical bases.

The modal properties of things include their causal powers, capacities, and propensities. For all such properties are dispositions to act on, or react to, other things, or to change spontaneously in some way. Therefore, on Hume's theory, the causal powers, capacities, and propensities of things must all have categorical bases. But these bases do not themselves determine behavior. It is the province of the laws of nature as they apply to things having these categorical properties to determine how they must behave.

The thesis that the causal powers, capacities, and propensities of things all depend on what categorical properties they have, and what regularities there are concerning their behavior, is sometimes called the Humean Supervenience Thesis.[9] If this thesis is correct, it implies that no dispositional or modal properties can exist as properties in their own right. The existence and nature of all such properties must be fully determined, once the categorical properties of things, and the ways in which things having only such properties behave, are fixed. For the theory implies that things can have the modal or dispositional properties they do only if they have the appropriate kinds of categorical bases, and the behavioral laws of nature are such as to ensure that things having such bases are disposed to behave in the appropriate ways.

The Humean Supervenience Thesis thus implies a radical distinction between properties that characterize things independently of their dispositions to behavior (categorical properties) and properties that determine how things would (or might) be disposed to behave in given circumstances (dispositional properties).[10] Although it is very widely accepted by philosophers, it will be argued here that the Humean Supervenience Thesis is untenable.[11] Indeed, we suppose that belief in Humean Supervenience has stood in the way of a satisfactory solution to the problem of natural necessity. For if modal properties do not exist in the actual world, it is hard to see how else natural necessities could be grounded. Either we must accept the whole Humean package, and deny their reality altogether, or we must find a basis in some other reality than actual reality. Astonishingly, many philosophers have chosen this latter option.[12]

Because categorical properties are supposed to be capable of existing independently of the laws of nature, we can perhaps envisage a world that is just like the actual world in all of its concrete detail, but does not have any laws of nature. Such a world is called a *Hume world,* and the world we have imagined is the Hume world that corresponds to this world. In this imagined world, there are no causal powers, capacities, or propensities: Everything that happens there appears to happen just as it does in this world, but instead of its happening of necessity, it happens just by chance.

Given the concept of a Hume world, we may redefine the categorical properties as those first-order properties that could be instantiated in a Hume world.[13] The paradigmatically categorical properties are spatio-temporal. That is, they are the kinds of properties that depend ontologically on how things are distributed in space and time. Shape is the obvious example of such a property. Other properties that are often said to be categorical have a kind of mixed status. That something has a certain crystalline structure, for example, depends on the spatial arrangements of the atomic or molecular units that make it up. To this extent, it is categorical. However, the property of being a crystal of a certain kind is not purely categorical. For this property is also ontologically dependent on the kinds of bonding between the atomic or molecular units and hence on the kinds of causal links that exist between them.[14] Such causal links, which explain the stability of the crystal structure, its cleavage planes, and so on, could not exist in a Hume world.

According to the Humean metaphysic, what is contingent is which, if any, universal regularities exist. Consequently, any changes of behavioral dispositions that could conceivably occur are in principle possible. A billiard ball, for example, might cease to behave as we expect, and instead become disposed to run around in circles when it is hit. Or our hands might become immune to flames. Such happenings cannot be ruled out as physically impossible, unless the laws that disallow them are somehow physically necessary. The problem for any Humean is therefore to make good sense of the claim that the laws of nature are both contingent and physically necessary.

Given this metaphysic, it is inevitable that dispositions should not be regarded as properties in their own right, but at best as properties that supervene on categorical properties and laws of nature – hence the dependent second-rate status that is commonly ascribed to them. It will be argued, however, that there are genuine dispositional properties that do not depend on the existence of underlying categorical structures and laws of nature, and that there is often a necessary connection between an object and the behavior it is disposed to manifest. It is not denied that things often have categorical properties. But there does not appear to be anything that only has categorical properties (although if there were, then I suppose we should not know about it).[15] In contrast, there are certain fundamental things in nature that appear to have *only* dispositional properties, and not to have any categorical properties at all. Their real essences, it would seem, are entirely dispositional. So it is plausible to suppose that the dispositional properties of such entities may be ontologically primitive.

1.10 EMPIRICIST AND ESSENTIALIST PERSPECTIVES ON REALITY

The Humean metaphysic, which is also the perspective of traditional empiricism, is deeply flawed. For the metaphysic implies that the causal powers and capacities of things are properties that depend on what the laws of nature are. If the laws were different, empiricists say, so would their causal powers be different. Moreover, since they suppose the laws of nature to be contingent, they must also suppose it to be a contingent matter what the causal powers and capacities of things are. Hence, they say, it must be a contingent matter how things are disposed to act or interact with each other, and consequently to impinge on our senses: In themselves, things are intrinsically powerless.

The things that exist in nature might have certain structures that dispose them to behave in certain ways, given that the laws of nature are what they are. But in another world, these laws might well be quite different, and things with exactly the same structures, composed of just the same substances, might be disposed to behave in other very different ways. Conversely, things with different structures, and made of different substances, might be disposed to behave in just the same way, if the laws of nature were sufficiently different.

The empiricist tradition thus draws a sharp distinction between the identity of a thing and its dispositions to behavior. What it is depends on what it is made of and how it is constituted. How it is disposed to behave depends also on what the laws of nature are. Consequently, the old Aristotelian idea that the identity of a thing might depend on its essential nature, which would dispose it to behave in certain ways, is firmly rejected by empiricists. The identity of a thing is supposed to be logically independent of what it is disposed to do, and hence independent of the perceptible qualities by which it may be known.

The scientific essentialist[16] view is that there are genuine causal powers, capacities, and propensities that are not properties that depend on what the laws of nature happen to be.[17] They exist in nature as universals, and are therefore the same in all possible worlds. And since these properties are the truthmakers for their laws of action, these laws must be the same in all possible worlds. Of course, there might be worlds in which these properties are not instantiated, or in which there are dispositional properties of other kinds. But if they are instantiated, the objects in which they are instantiated must, by virtue of having these properties, be disposed to behave as the laws of action of these properties determine. Consequently, if there are any things belonging to natural kinds that have such properties as these necessarily, these things must be disposed to behave in the required manner, just because they are things of the kinds they are.

To the extent therefore that the laws of nature describe the behavioral dispositions that things must have in virtue of being things of the natural kinds they are, they are necessary. To illustrate: If a and b are electrons, then it is necessarily true that they are negatively charged. And, necessarily, if a and b are negatively charged, they are intrinsically disposed to repel each other. Therefore, if a and b are electrons, it is necessarily true that they are intrinsically disposed to repel each other. It is therefore a necessary truth, not a contingent one, that electrons are intrinsically disposed to repel each other. The law that electrons are intrinsically disposed to repel each other is therefore true in all possible worlds. It is non-vacuously true in every possible world in which electrons exist.

What makes something an electron, a scientific essentialist would say, does not depend on the shape, size, number, or material constitution of its primary parts, as a Humean would have to suppose. It depends on what its causal powers, capacities, and propensities are. For anything to be an electron, it must have a certain mass, charge, spin, stability, and so on and therefore be capable of acting on, or interacting with, other particles and fields in certain ways, depending on the laws of action of these properties, and the powers, capacities, and propensities of the things on which it acts or which act on it. If anything lacked any of these causal powers, capacities, or propensities, it would not be an electron. Unit charge, unit mass, and spin 1/2 are essential properties of electrons, and electrons are by their very nature bound to act and interact as these properties determine.

1.11 CAUSAL POWERS AND CAUSAL PROCESSES

The metaphysical position of scientific essentialism, which is the perspective from which this book is written, is thus very different from the Humean position. Scientific essentialists are realists about the intrinsic causal powers, capacities, and propensities of things. They believe in them, not just in a manner of speaking, but as genuine occurrent properties (Ellis and Lierse, 1994). They also think that in at least some cases, such properties may be ontologically basic – that is, not dependent for their existence on other properties. Scientific essentialists therefore reject the categorical realists' claim that such dynamical properties as these, which are dispositional in character, are necessarily ontologically dependent on other properties that are not dispositional. That is, they would deny that the causal powers, capacities, and propensities of things must be ontologically dependent on categorical properties (such as number, shape, size, and configuration of parts) of things and the laws of nature (Ellis and Lierse, 1994). The Humean (and Lockean) picture of an intrinsically inert world governed by contingent laws of nature, which is the philosophical basis of categorical realism, is rejected.

Scientific essentialists are also realists about the entities they postulate as constituents of things, as well as about the structures they suppose to be essential to the natural kinds of objects whose properties and behavior they are seeking to explain. Thus, if one natural kind of thing is a constituent of another, then it is so necessarily. For example, since hydrogen and oxygen nuclei are constituents of water, water necessarily contains both hydrogen and oxygen nuclei as constituents. A substance would not be water if it did not have these constituents. But not everything with hydrogen and oxygen nuclei as constituents is water. To be water, a substance must also have a certain molecular structure consisting of two hydrogen nuclei and one oxygen nucleus, and these nuclei must be bound together in a certain way to form molecules with a certain electron structure. If a substance lacked any of these features, it would not be water. What makes a substance water, therefore, depends on both its constituents and its structure. These it has necessarily.

Consider the case of two isomers, one of which rotates the plane of polarization of light in one direction, while the other rotates it in the opposite direction. To explain this difference, it may be supposed that they have molecular structures that are mirror images of each other. Thus, if one of these structures – call it the "right-handed" structure – explains why one of the isomers is dextro-rotatory, while its mirror image explains why the other structure – call it the "left-handed" structure – while chemically very similar, is levo-rotatory, then we have potentially a very good explanation of the salient difference between them. And any scientific essentialist who accepts the explanation as adequate must be a realist about the molecular structures that are postulated in this explanation. Otherwise, he has no explanation at all of the kind that is being sought. For a satisfactory essentialist explanation, the observable difference in the direction in which polarized light is rotated has to be explained by reference to an essential difference between the underlying molecular structures. A scientific essentialist must therefore be a realist about the theoretical entities and structures postulated in such explanations.

But perhaps what is most distinctive of the scientific version of essentialism is that scientific essentialists are realists about natural kinds of processes, as well as natural kinds of objects and substances. Moreover, the natural kinds of processes that do exist are supposed to have essential natures – that is, causal or stochastic structures that make them the kinds of processes they are. Consider refraction. For anything to be a case of refraction, it must occur in a certain way. It is not enough that it look like a case of refraction. It is not even enough that it obey the manifest laws of refraction. One can imagine – although I do not know whether it is really possible – that a phenomenon of mock-refraction might be produced using hollow glass prisms and concealed diffraction gratings, so that the diffracted light emerging from the prism does so in precisely the way it would if the prism had been solid and had the refractivity of glass. The realism of the example is irrelevant here. What is important is that however similar the mock refraction may be to real refraction, it would not be a case of refraction.

For anything to be a case of refraction, it has to be a display of refractivity. The kind of mock refraction just envisaged is not a display of the refractivity of any of the substances involved in the construction of the apparatus. Therefore it is not a case of refraction. Precisely what refraction is is not easy to say. There are quantum mechanical theories about the transmission of light through a refracting medium that probably go some way toward describing the essential nature of this process. But exactly what the process is may not yet be fully understood.

Not all causal processes can have structures of the sort that refraction presumably has, unless there are some actually infinite regresses of causal processes going "all the way down." Sooner or later, in the process of ontological reduction, we must come to events and processes that are not themselves structures of constituent causal processes. These most elementary causal processes, I suppose, will consist entirely of elementary events (for example, basic causal interactions between particles) and certain acausal (for example, inertial) energy transmission processes which are initiated, and eventually terminated, by or in such elementary events. An event of radioactive decay, for example, may produce a β-particle which carries energy (and also spin, momentum, and other conserved quantities) to eventually interact with, and produce a reading, on a Geiger counter.

This may serve as a model for elementary causal processes generally. The first event (of radioactive decay) is a species of those events I am calling "causal interactions." They are instantaneous, as far as we know, and have no describable stages. The next stage in the causal process is the transmission of energy from the initiating event (the cause) to the terminating event (the effect). This is an inertial process governed by the quantum-mechanical laws of particle transmission. Qua inertial process, energy transmission requires no cause to sustain it, and it will continue indefinitely or until it is terminated by another causal interaction. The final stage in the causal process (the effect) is the interaction that occurs in the Geiger counter. This is another of these instantaneous causal interactions that has no describable stages. The cause and effect events are sudden discontinuous changes of state which occur according to the conservation and other laws governing particle decays and interactions.

The events and processes constituent of causal processes of any given kind must themselves belong to natural kinds. For causal processes of the same kind must be initiated by causes of the same kind, be carried by energy transmissions of the same kind, and terminate in effects of the same kind. That is, the different causal processes that instantiate any given natural kind of causal process must all have the same causal structure. If they had structures of different kinds, they would not be causal processes of the same kind. However, the identities of the basic causal interactions that initiate and terminate elementary causal processes, and the energy transmission processes that connect them, cannot depend in turn on their causal structures. For, by hypothesis, they have no causal structures. Elementary causal processes have a structure of some kind, for there are laws of causal interaction, as well as laws of energy transmission. But these are not, in my view, laws that describe causal processes. Schrodinger's equation and Maxwell's laws of electromagnetism are not causal laws, but the laws of energy transmission on which the laws of refraction and other causal laws ultimately depend.

51

Whether this analysis of causal processes, showing them to be ultimately dependent on instantaneous causal interactions and temporally extended acausal energy transmissions, is finally satisfactory, is not for me to decide. It is for science to discover what kinds of events and processes are foundational. But some such ontology is needed if we are to break away from the grip of Humeanism. On a Humean ontology, there are just instantaneous events located in space and time. These are thought to be independently classifiable as events of various kinds, and the causal laws are supposed to describe the ways in which the localized events of these various kinds succeed each other. Given this picture, it is hard to see how the laws could be anything but contingent.

However, I am not inclined to start with this picture. It is not the view which modern science requires us to take. The quantum-mechanical view of nature is much more holistic. The Humean ontology also presupposes that the primitive events that occur in nature have kind-identities that are independent of their causal roles. I do not believe this – and this book is largely an attempt to explain why I don't believe it. For a dynamic ontology of the sort required for the kind of world that science is revealing to us, one has to assume that there are fundamental processes of various kinds that are not just sequences of instantaneous point events whose identities are independent of the processes in which they are involved. There is a holism about these processes that makes it impossible to analyze them in this way. So one has to think of them as processes essentially described by the laws of their development.

What is essential to any acausal energy transmission process is presumably the law or laws of its propagation. An electromagnetic wave must be transmitted according to laws of electromagnetism. Necessarily, energy that is transmitted in any other way is not carried just by electromagnetic radiation. Likewise, a particle of matter cannot be transmitted inertially from one place to another except according to the specific wave-mechanical laws for such energy transmissions. (For macroscopic objects, these laws, and the Newtonian laws of inertial motion, yield indistinguishable predictions.) I therefore postulate that these laws are of the essence of these processes. They are what make these processes the kinds of processes they are.

1.12 CAUSAL POWERS AND CAUSAL LAWS

Realism about primitive causal powers, and an essentialist theory that attributes such powers to fundamental natural kinds, implies the necessity of at least some causal laws involving things of these kinds. For the identity of any causal power depends on what it is disposed to do. Therefore, things that have the same causal powers must be disposed to exert the same forces on things of the same kinds in any worlds in which they might exist. There can be no question of a causal power's acting one way in one world and in a another way in a different world. Therefore, if any causal powers are primitive – that is, not dependent on other properties or laws of nature for their existence, then things that have these powers necessarily must be disposed to behave in the same sorts of ways in any world in which they might exist.[18]

To illustrate: Electrons have unit mass and unit charge necessarily. Therefore, electrons must be disposed to behave as particles with unit mass and unit charge in any world in which they might exist.[19] A particle with unit mass has, by virtue of this property, a certain power to act gravitationally, and to resist acceleration by a given force, while a particle with unit charge necessarily has a certain power to generate electromagnetic waves and to respond to them in certain ways. Therefore, in any world in which electrons might exist, the fields that carry such forces or waves must also exist. The identities of these fields themselves depend on their causal powers and capacities. The identity of a field depends on how it is transmitted, what is affected by it, and how it is affected. Therefore, in any world in which electrons might exist, there must be gravitational and electromagnetic fields, and electrons must be disposed to respond to these fields in essentially the same ways as they do in this world. The causal laws concerning interactions between electrons must therefore hold in any world in which electrons exist. Hence, they are metaphysically necessary.

On the view taken here, all of the basic causal laws are necessary in this sense, and for the same kinds of reasons. It is not of course the case that these laws are knowable *a priori*. We have to *discover* what the essential properties of the fundamental natural kinds are and to what physical behavior they dispose their bearers. Ours is not an *a priori* essentialism. We think that the laws concerning the behavior of the most fundamental kinds of things in nature are *a posteriori* necessary.[20]

The main difficulty that most people see with this position is its counter-intuitiveness. Intuitively, most people want to say, it is a contingent fact that like charges repel each other. Yet, if what I am saying here is correct, this is not a contingent fact, but a real or metaphysical necessity. I say unequivocally that it is a metaphysical necessity, and if we are inclined sometimes to think that the law could have been otherwise, it is because we are not very good judges of metaphysical possibility and we are inclined to think that whatever is conceivable is also really possible. I dispute this. Conceivability is not a good test for real possibility. What is conceivable depends on our mental capacities. What is really possible depends on the kind of world we inhabit.

Thirty years ago, many Anglo-American philosophers, and probably most Australian philosophers, believed in contingent identities. Sensations, it was said, are contingently identical with brain processes – water is contingently identical with H_2O, the Morning Star is contingently identical with the Evening Star, temperature is contingently identical with average molecular kinetic energy, and so on. Saul Kripke argued, correctly in my view, that the concept of contingent identity is an oxymoron. If $a = b$, then this relationship holds necessarily, not contingently. To say it holds contingently is to confuse contingency with *aposteriority*. Certainly, none of these identities can be known *a priori* to exist. They can be known only as a consequence of scientific investigation. But this does not make these identities contingent. All it shows is that some of the results of scientific investigation are discoveries of necessary truths, rather than contingent ones. Of course, it is possible for the word "water" to be used to designate something other than H_2O, or for "the Evening Star" to refer to some object different from the Morning Star. After all, it is contingent how words are used. But if it is true that the object of one such naming or referring expression (a "rigid designator" in Kripke's terminology) is the same as that of another, the statement asserting the identity of these objects of reference is not only true, it is necessarily true.

We need not go into the details of Kripke's theory. There is a huge volume of literature on the topic, and it is not my intention to add to it. It is important to note, however, that my concerns are different from Kripke's. Kripke's essentialism was developed in relation to theories of reference and identity. Scientific essentialism derives from an examination of the scientific practice of theoretical identification. Kripke's essentialism has a lot to say about individual essences. Scientific essentialism is almost exclusively concerned with the essences of natural properties and the kind essences of things, and has nothing new to say about what makes an individual thing the individual thing it is. Scientific essentialism is primarily concerned with the question of what makes a thing the kind of thing it is, and so display the manifest properties and behavior it does. That is, real essences for us must not only be identifying, they must be explanatory.

The real essence of any natural kind is a set of properties or structures in virtue of which a thing is a thing of this kind, and displays the manifest properties it does. It is not just any old set of identifying properties – that is, properties that are severally necessary and jointly sufficient for membership of the kind. The essential properties of a kind include all of the intrinsic properties and structures that together make a thing the kind of thing it is.

The electron, as far as I know, is the only particle with unit mass and unit negative charge. It is also, as far as I know, the only stable, negatively charged particle with spin 1/2. In virtue of what, then, should we say that a particle is an electron? On my view, all of these properties are among essential properties of electrons, because a particle would not behave like an electron, and hence not be an electron, if it lacked any of them. To be an electron, a particle must behave as electrons do. And it can only do this if it has all of the standing properties of electrons, not just some identifying subset of such properties.

The scientific quest to discover the real essences of the various natural kinds is therefore not just a search for their identifying properties, structures, or ways of behaving. The scientific task is to discover what makes a thing the kind of thing it is and hence to explain why it behaves or has the properties it has. The scientific version of essentialism is therefore less concerned with questions of identity, and more with questions of explanation, than is the classical essentialism of Aristotle or the new essentialism of Kripke. Its closest historical predecessor is the kind of essentialism described by Locke. For Locke too was concerned with the question of what makes a thing the kind of thing it is. He thought that if only we knew this, we should be able to explain why it has the manifest properties it has and behaves as it does. But Locke despaired of discovering such knowledge, and did not think we should ever be able to attain to it.

However, Locke was unduly pessimistic. The basic sciences of physics, chemistry, and microbiology have now revealed many of the intrinsic properties and structures of the things that exist in nature, and we are now justifiably able to say what makes them the kinds of things they are and why they are disposed to behave and display the properties they do. In Locke's day, the underlying causes of most ordinary phenomena were unknown, and apparently unknowable. Today, many of these causes are known, and we are well justified in being scientific essentialists and hence embracing scientific realism.

1. Hacking (1991) distinguishes four principles concerning natural kinds. (1) *Independence:* They exist independently of psychological and social facts about human beings. (2) *Definability:* Natural kinds are definable. (3) *Utility:* Recognition and use of natural kinds plays a significant role in the development of human knowledge.

(4) *Uniqueness:* There is a unique best taxonomy in terms of natural kinds. Hacking rejects (4), but argues that kinds that fail to satisfy the first three conditions "should not be assimilated to the conception of natural kinds however "natural" the categories they correspond to may otherwise be" (Boyd, 1991, 128). De Sousa (1984) argues that "'the philosophical search for natural kinds is motivated by the hope of finding ontological categories that are independent of our interests. But no such categories are to be found. Virtually any kind can be termed 'natural' relative to some set of interests and epistemic priorities" (p. 562). In a section entitled "Seven Characteristic Features," De Sousa lists eight characteristics he thinks natural kinds should have: objectivity, explanatory primacy, multiplicity of kinds, sharp boundaries, stability, uniqueness of membership, equal stability, perspicuity. Of the natural kinds that are commonly supposed to exist, De Sousa argues that their objecivity is a mere presumption, their explanatory role can be more or less basic, they are multiple but so numerous as to be trivially so, their boundaries are not always sharp, they are unevenly stable, they are not unique in membership nor equipollent, and they are the less perspicuous the deeper in the hierarchy they lodge (p. 580).

2. Thomason (1969) argued that any natural classification system must have the structure of an *upper semi-lattice.* That is, he took the view that if anything is a member of more than one natural kind, then one of these kinds must be a species of the other. However, this is a stronger requirement than is needed, and does not account for the structure of chemical kinds, which we are taking as our guide. The natural kinds *cupric compounds* and *sulfates* have a common member – copper sulfate – but neither is a species of the other

3. This point has been elaborated in Ellis (forthcoming).

4. The color-kind *red* is not strictly a natural property kind, according to my criteria. However, this is irrelevant to the point being made.

5. Kim's (1982) definition is actually a temporally constrained version of this. However, it is neither necessary nor desirable to constrain the definition of internality in this way.

6. The concept of a differential force is due to Hans Reichenbach. Reichenbach used this concept to define a rigid body: "Rigid bodies are solid bodies which are not affected by differential forces, or concerning which the influence of differential forces has been eliminated by corrections; universal forces are disregarded" (Reichenbach, 1958, 22). Thus, as defined by Reichenbach, a rigid body is a theoretical ideal. It is simply any body – ideal or actual – whose shape is just its intrinsic shape.

7. David Lewis (1983a, 197) has described such a circle.

8. This is perhaps the main reason why Locke doubted the existence of natural kinds. Ayers (1981) comments on Locke on real essences:"... [Locke] holds that what sets a boundary to [a] class is always what he calls the 'nominal essence,' i.e., the abstract idea that embodies our criteria for the application of the kind-name or sortal. What explains the properties of the species so defined, on the other hand, is corpuscularian structure (or at least something like it, if Boyle's theory is less than the whole truth). Those aspects of the structure of the individual members of a species which they have in common and in virtue of thich they all possess the defining properties of the species comprise what Locke calls the 'real essence' or 'constitution' of the species. The distinction between nominal and real essence, derives, of course

from the Aristotelian distinction between nominal and real definition" (p. 256). Ayers adds that "... to say as [Locke] does that members of a species defined by a nominal essence have, or probably have, corresponding similarities at real essence-level – so that we can talk of a corpuscular real essence of the species – is not to concede that the real essence could, independently of any human decision, determine ontologically the *boundaries* of the species. Locke is simply saying that, if we select some set of observable qualities to serve as our nominal essence, then no doubt behind this arbitrarily selected phenomenal resemblance will lie a structural resemblance indirectly picked out by the same arbitrary procedure of selection. He explicitly makes the point that, even if we knew the 'real essence' in this sense, all the problems about boundaries would rise up again: "For what is sufficient in the inward contrivance to make a new species?" (p. 259).

9. Carroll (1994) discusses a number of Humean supervenience theses, including the thesis to which regularity theorists often retreat – that if two worlds agree on all non-modal facts, then they must also agree on all modal facts. He produces a number of good examples (see Chapter 3) intended to show that these theses are all untenable.

10. The distinction between categorical and dispositional properties is an important one. It will be discussed in more detail in Chapter 3 at Sections 3.5, 3.6, and 3.12.

11. In rejecting the Humean Supervenience thesis, I share common ground with F.I. Dretske, M. Tooley, D.M. Armstrong, J.W. Carroll, and a number of other philosophers who have become dissatisfied with the Humean paradigm.

12. See Lewis (1986b) and Bigelow and Pargetter (1990).

13. See Bigelow and Pargetter (1990), pp. 238–245, for an account of the Hume world.

14. For example, if the molecules in a glass of water were by extraordinary chance to become momentarily arranged in a cubic lattice, the water would not for that moment become crystalline. To become crystalline, the water would have to freeze and the appropriate kind of bonding between the molecules would have to be established.

15. The question of the status of categorical properties will be taken up again later in response to objections by Swinburne and Armstrong (see Section 3.12).

16. "Scientific essentialism" is the name that both George Bealer (1987) and I (1992) have independently given to the view that explanations in the physical and biological sciences typically involve postulates concerning the essential natures of the fundamental natural kinds of objects and processes occurring in the world. On this view, the causal laws that apply in these areas describe the operation of the essential properties of these natural kinds, and are metaphysically necessary.

17. Mackie (1977) wholly rejects the claim that there are intrinsic causal powers in things. He labels this 'the rationalist' view, and parodies it using the hypothesis that sleeping drafts have dormative virtues. He says: "Such intrinsic powers are pretty clearly products of metaphysical double vision: they just *are* the causal processes which they are supposed to explain seen over again as somehow latent in the things that enter into these processes (p. 104).

18. On Achinstein's (1974) theory, property identity generally depends crucially on causal role: If P and Q are the same property then anything that has P necessarily has the same causal role as anything that has Q. Of course, this is putting it very crudely, and Achinstein is at pains to spell out this idea very carefully. His thesis

VIII is that if $P = Q$, then for all x and e, it is necessarily true that x's having P is causally relevant for e (or conversely) iff x's having Q is causally relevant for e (or conversely). The concept of causal relevance that is used in this definition is in turn defined by a number of sub-theses. Together, these sub-theses define what he is calling "causal role." (Note that the necessity operator in thesis VIII occurs within the quantifiers, and is therefore a *de re* necessity operator.) However, Achinstein's thesis concerning property identity is stronger than the one that is required here.

19. Averill (1990) argues that the property of inertial mass, which is said to measure a body's capacity to resist acceleration and therefore almost paradigmatically dispositional, is not dispositional at all. It is not dispositional, he argues, because it does not entail the required counterfactual conditionals. However, Averill's arguments are unsound, as Reeder (1995) adequately shows.

20. Freddoso (1986, 233–4) argues that it is "impossible that there should have been different laws of nature if by this we mean either (i) that the substances that in fact exist could have had different natures, i.e., could have instantiated different natural kinds, or (ii) that the natural kinds instantiated by these substances could have been necessarily tied to basic causal dispositions different from those they are, in fact, necessarily tied to. Nonetheless, there could have been different laws of nature in the sense that it is metaphysically possible (i) that there should exist natural substances of kinds that have, in fact, never been instantiated and/or (ii) that the natural kinds that are, in fact, instantiated should not have been. So, for instance, my position does not rule out the possibility that there should have been elementary physical particles of types that are different from those that are, in fact, instantiated and that are such that if they had been instantiated instead, then the strengths of the fundamental forces in nature would be slightly different or even greatly different from what they are in our universe. Under those circumstances, the scientific theories that hold of our universe would not have held" (pp. 233–4).

Part Two

Ontology

2

Natural Kinds

2.1 THE AIMS OF ONTOLOGY

An ontology is a theory about the basic constitution of the world that seeks to explain some of its more general features. It is therefore likely to depend on what we take the general character of reality to be. For example, one might reasonably expect a contemporary ontology to explain:

1. Why the world is lawlike in its behavior and not just a motley collection of objects and events (or of states of affairs) bearing little relation to one another.
2. Why the world appears to have a hierarchical structure of natural kinds of objects (substantive kinds), processes (dynamic kinds), and properties or relations (property kinds).
3. What the laws of nature are and how they are grounded in reality.
4. The importance and role of quantities in the laws of nature.
5. The existence of a hierarchy of laws that nature, ranging from very general laws which apply non-vacuously to all events and processes to quite specific laws that apply non-vacuously only to quite specific kinds of things or processes.
6. The nature of natural necessity, and how it is grounded in the world, if indeed it is so grounded.
7. The dynamism of the world – that is, the fact that the world seems to be made up of things constantly interacting with each other.

There are of course many other general features of reality besides these – for example, the overall physical and space-time structure of the world. But the features listed here are certainly some things that one might reasonably expect an ontology to deal with.

Ontologies typically try to explain the overall structure of the world, not by deriving its salient features from any more general characteristics, for that would only raise further questions of the same kind, but by presenting a simple and coherent world view that incorporates them. A good ontology should therefore seek to establish connections between such facts as these, thus enabling them to be seen in proper perspective, with each general fact falling naturally into place. All explanation involves establishing plausible links between things that would otherwise be dissociated. But ontology takes on the really big questions. It aims to link together those general features of the world that at present seem disparate and inexplicable. The test of an ontology is how well it achieves its aim of global unification.

The questions asked by philosophers are not the same as those asked by physicists. Physicists might wonder what the ultimate constitution of the world is, and speculate about various possibilities. Can the various particles and fields that we know about be explained in terms of things that are more fundamental? Or, can the particles be reduced somehow to fields in a complex and convoluted space-time? Philosophers, on the other hand, are not so much concerned with what specific kinds of things might be supposed to exist at the most fundamental level. If it is agreed that the most fundamental things in nature are physical entities of some sort – things with energy, acting and interacting with other physical things, according to their various natures – then the task of identifying these things and their natures can be left to the physicists. The philosopher might well doubt whether the most fundamental existents are things of any sort.

Wittgenstein and Russell, for example, held that the world consists of facts, not things. Armstrong (1997), whose philosophy belongs to the same tradition, thinks that the world is made up of states of affairs. The states of affairs existing at a given time are roughly the states of things existing at that time and the relationships that then exist between them. So states of affairs are static images of a sort, like the frames of a moving picture. The world, to continue the analogy, is the total picture that results as the frames succeed one another in temporal order.

Whether this is a satisfactory basis for an ontology is not a matter for immediate judgment. There are difficulties with every ontology, including the more intuitive one of things with attributes. The test of an ontology is two-fold: First, it must be defensible against enlightened criticism concerning the natures of, and relations between, the elements of the ontology, and second, it must succeed in its tasks of ontological reduction and conceptual unification. On the whole, I think that Armstrong does a good job of defending his ontology against detailed criticism, but judged globally as a theory of reality, it is unconvincing (See Ellis, 1999d).

The basic concept on which any ontology must depend is that of onto-logical dependence. For if it can be established that As are ontologically de-pendent on Bs, but Bs are not ontologically dependent on As, then Bs must be ontologically more fundamental. To illustrate: States of affairs can be more or less elementary. Whether there are any truly elementary or atomic states of affairs is moot, but it seems clear that some states of affairs depend on others. A's being a glass B of water C, for example, depends on B's be-ing a glass, C's being water, and C's being in B. We say that the first of these states of affairs depends ontologically on the three that follow, because the first state of affairs could not exist if the others did not exist. Armstrong postulates that if we thus follow down the channels of ontological depend-ence of states of affairs on states of affairs, we should eventually arrive at certain atomic states of affairs on which all others must ultimately depend (although he concedes that this process might have no end).

Given a satisfactory theory of ontological dependence (see Section 2.6), I have no objection to the program of trying to discover what is on-tologically most fundamental. On the contrary, I hold that this is what ontology should be all about. However, I begin with the assumption that it is most rational to start from a scientifically well-informed view of what there is and how the world is structured, and to work back from there. One should not, for example, proceed *a priori* to posit a global view of the nature of reality and then seek to reconcile this view with the scientific one. For the questions we have to ask are: What do we have to believe in if we wish to accept the scientific world-view, and how can we best ex-plain the presently unexplained general features of the world according to this view?

Armstrong (1997) begins from the Tractarian position that the deep structure of reality is propositional, and so is really a vast complex of states of affairs. His main problems are then to identify and characterize the most elementary states of affairs and explain how the more complex ones are constituted by the more elementary. Given this conception of reality, one would expect the relation of ontological dependence to mirror that of log-ical dependence. That is, one would expect the criterion for ontological dependence among properties to be: A property A is ontologically de-pendent on a property B iff nothing could have the property A unless something has the property B.

However, it is not clear that this is how Armstrong thinks about it. Arm-strong employs a concept of supervenience to settle questions of ontological dependence. Thus he says that if one thing is supervenient on another, then its existence requires nothing additional to the existence of that other. There is, as he puts it, "no increase in being" involved. But how are we to judge what is meant by "an increase in being"? My intuitions and his are evidently at odds on this point. Presumably the aim is to discover what is ontologically dependent on what, and so, finally, what exists most fundamentally. That being the case, the relation of supervenience must be intended to be the equivalent of that of total ontological dependence. That is, A supervenes on B, iff A is wholly ontologically dependent on B.

The accepted paradigm of ontological dependence is to be found in the theory of micro-reduction. Methane molecules, for example, are said to depend ontologically on their constituent hydrogen and carbon atoms. They are said to be ontologically dependent because the methane molecules could not in fact exist if these atoms did not exist. Conversely, however, the atoms could exist, even though the molecules did not exist. Let us put this another way by saying that methane molecules existentially entail hydrogen and carbon atoms, but hydrogen and carbon atoms do not existentially entail methane molecules. The direction of ontological dependence (from methane molecules to hydrogen and carbon atoms) is therefore the same as the direction of existential entailment. That is, As are ontologically dependent upon Bs iff the existence of As entails the existence of Bs, but the existence of Bs does not entail the existence of As.

But in Armstrong's (1997) theory of supervenience, as it applies to properties and kinds, the opposite seems to be the case. "A point to be taken very seriously," he says, "is that because determinates entail the corresponding determinable, the determinable supervenes on the determinates, and so, apparently, is not something more than the determinates" (p. 50). Armstrong's argument for this conclusion is that if we know that Bs exist and that the existence of Bs entails the existence of As, then we add nothing new by postulating As over and above Bs. There is no such thing, he says, as an "ontological free lunch." To illustrate: The existence of a determinate quantity entails the existence of the determinable (the quantity itself of which it is a determination), but the existence of the determinable does not entail the existence of the determinate (although the quantity, and hence, on Aristotelian assumptions, some determinate value of the quantity, must be instantiated). Therefore, he says, the determinable quantity is ontologically dependent on its determinate values. So if Armstrong is right, the direction of ontological dependence among properties and kinds (from determinables to determinates) is contrary to the direction of existential entailment.

Perhaps Armstrong is right about this. Perhaps the direction of ontological dependence among kinds of substances is one way (that is, the same as that of existential entailment), while the direction of ontological dependence among properties is the other way (that is, opposite to the direction of existential entailment). Or perhaps, as he later seems to suggest (at 16.13), that genuine quantities (for example, those involved in laws of nature) might be universals in their own right, and are not supervenient on the universals that are their specific values. But I do not see how Armstrong can possibly maintain either of these theses. For the "properties" of being a methane molecule, being a hydrogen atom, and being a carbon atom, are regarded by Armstrong as a complex properties, even though he has some doubts about their status as properties rather than kinds (p. 35). Therefore, for Armstrong, ontological dependencies among substances must run in the same direction as ontological dependencies among properties.

However, the direction of ontological dependence has to be the same as that of existential entailment. It cannot be the other way around. To reinforce this point, consider shape. The existence of triangular configurations is surely ontologically dependent on the existence of spatial relationships. For the existence of triangular configurations entails the existence of spatial relationships, but the existence of spatial relationships does not entail the existence of triangular configurations. The direction of ontological dependence is therefore the same as the direction of existential entailment in the case of shape, just as it is in the paradigm cases of ontological dependence taken from chemistry. The conclusion therefore seems to be that Armstrong is simply wrong about the direction of ontological dependence among properties. The more general property kinds (quantities and other determinables) are ontologically more fundamental than the more specific.

There are indeed some independent reasons for believing this to be so, the main one being that quantities cannot be constituted by their values. Most quantities have gaps in their spectra that are, and probably never will be, instantiated. For example, there is probably no object anywhere in the universe with a mass equal to $m/2$, where m is the mass of the electron. Therefore the property of having such a mass is probably an uninstantiated universal, and so does not exist in our world. Therefore a quantity constituted by a set of universals of the kind belonging to the determinable *mass* would probably not be a continuous one in our world. In most other possible worlds, however, the gaps would almost certainly not be the same as in our world. Therefore, we should have to conclude that the quantity mass, as it is constituted in this world, does not exist in most other possible worlds. We should also have to conclude that there is no possible world in which anything of mass $m/2$ exists (since the determinable in this other world, whatever it might be, would not be *mass*). Therefore, any counterfactual conditionals speculating about how particles with mass $m/2$ would have to behave could only be vacuously true.

To solve this problem, there are really only two ways to go. One is to allow that quantities are ontologically at least as fundamental as any of their infimic species, and to admit that there are generic as well as classical universals. The other is go upward to second- and higher-order universals, and to argue that quantities and other determinables are at least second-order. The first alternative has some difficulties associated with it, that are not insurmountable. But it does require a clear distinction between *infimic species* and *specific instances* (see Section 1.3). The specific quantitative property of having mass m, for example, must be considered to be an infimic species of the genus, having mass, and therefore not a specific instance of it. A specific instance of the quantity mass would be any trope of any infimic species of the generic property of having mass. An instance of the specific quantitative property of having mass m, on the other hand, would have to be a trope of the property of having the specific mass m.

The other alternative is the second-order route taken by Bigelow and Pargetter (1988). It identifies the instances of a determinable D with its various determinations, d_1, d_2, d_3, ..., which are themselves universals. But this route is fraught with difficulties. Consider the sequence {plane figure, polygon, triangle, scalene triangle, scalene right-angled triangle, triangle of sides in ratio 3:4:5, triangle of sides 3m, 4m, 5m}. Each member of the sequence, apart from the first, is a species of the previous member. The last member of the sequence is a universal, which has no species, although it may have instances: It is an infimic species. Therefore the instance and species relations are clearly different. But the second-order route depends on denying this distinction. It treats triangle of sides 3m, 4m, 5m as a universal which is an instance of the second-order universal triangle of sides in ratio 3:4:5, which in turn is an instance of a third-order universal scalene right-angled triangle, which in turn is .... Hence, if no distinction is drawn between instances and species, then the first member of the given sequence must be at least a seventh-order universal (depending on the hierarchy of determinables accepted as universals) and each successive member must be both an instance and a species of the preceding member.

The membership/class-inclusion distinction would seem to be far too important to all of our thinking about the world to be abandoned just in order to avoid having to admit the existence of generic universals. Indeed, the paradoxes of set theory demand that this distinction be rigorously maintained. The cost of taking the second-order route that carries us onward and upward to higher-order universals is therefore much too high. In my view, quantities are generic universals, which are like classical universals, except that they permit a range of intrinsic variation other than zero. Specific quantitative properties are the ordinary classical universals that are the infimic species of quantities. Therefore, any instance of any quantitatively specific property is an instance of the generic quantitative property of which the quantitatively specific property is an infimic species, but not vice versa.

2.3 THE NATURAL KINDS STRUCTURE OF THE WORLD

Natural kinds are among the primary objects of investigation in the natural sciences. Chemistry alone demonstrates this. So some account of natural kinds has to be given if we are to construct an ontology that is adequate for the world as we understand it. In this chapter, I begin developing such an ontology. I do not say that the ontology to be outlined is the most economical that can be defended. But it is a start in the right direction. If it provides an adequate basis for explaining the general features of the world that any contemporary ontology should be able to account for, then the work of further ontological reduction can be left to others.

The most salient general feature of reality I seek to explain is its natural kinds structure. There are hierarchies of objects of increasingly complex kinds. There are hierarchies of events and processes, the more complex of which may be analyzed in terms of simpler, more elementary, events and processes. And there are hierarchies of properties and structures. The more complex structures are compounded of simpler structures, the more complex properties, of simpler ones. Within each of these hierarchies, we may detect various samenesses (exact resemblances). Two objects can be intrinsically identical. Two events or processes can be of exactly the same kind. Two property instances can be precisely the same, or two structures identical. These identities can best be explained by assuming that there are universals that have particular objects, particular events, or the tropes of specific properties, relations, or structures as their instances. I call these universals "substantive," "dynamic," and "property" universals, respectively.

In addition to these samenesses, there are also likenesses, which are less than perfect. Two objects can be of the same generic kind, being intrinsically identical in some respects, but not in all. Two events or processes can have the same generic structure without being specifically identical. Or two property-instances or structure-instances can be quantitatively different, or different in other ways, but nevertheless instances of the same quantities or generic structures. These similarities can best be explained by assuming that there are hierarchies of generic universals,[1] the infimic species of which are all specific universals with identical instances.

An ontology in which the only property universals that are recognized as fundamental are non-structural ones does not seem to be up to the task of accounting for the complex structures that are to be found in nature. This is true, even if we allow that there are plenty of non-structural universals of various adicities and degrees. The reason is simply that universals are not localizable in either space or time, but structures are necessarily spatial, temporal, or spatio-temporal.

In their theory of structural universals, Bigelow and Pargetter (1989) make a gallant attempt to reduce structural universals to non-structural ones. They focus on the alleged property of being a methane molecule. In my view, "being a methane molecule" is not a property name, but a predicate that is constructed out of a natural kind name, and so pretends to name a property (see Section 2.8). Be that as it may, there is a definite structure that is exhibited by all methane molecules, and hence a structural universal that is instantiated in all such molecules. The question is: Can this structural universal be reduced to some complex of non-structural universals? I think not.

According to Bigelow and Pargetter, the property of being a methane molecule is roughly equivalent to that of being a thing which consists only of one part that instantiates "carbon atom" and four parts that instantiate "hydrogen atom," each of which bears a certain relationship of "bonding" to the carbon atom, a relationship which no hydrogen atom bears to any other. But, at best, this accounts only for the molecular formula of methane, which is arguably non-structural. It does not account for the shape of the methane molecule, which is what is required. It cannot account for the shape for the simple reason that universals cannot be spatially arranged. Objects can be spatially arranged. But as soon as we speak of a spatial arrangement of objects, we are already invoking the language of structural universals, since different groups of objects can presumably

68

be arranged in exactly the same ways. I do not, of course, deny that there can be ontological dependencies between structural universals. Some structures are ontologically dependent on others. But at some point, one must come to fundamental structures of a spatial or temporal nature that cannot be further reduced.

To describe a methane molecule, it is necessary to say how the atoms are arranged within the molecule. The molecular formula is not enough. This becomes obvious if we consider molecules such as butane and iso-butane, or pentane, neo-pentane, and iso-pentane, which are distinguished from each other intrinsically only by their different molecular structures. Bigelow and Pargetter note this, but then dismiss the problem, saying that it will need some more work. I think that the problem is not one that can be overcome with more work if their aim is to reduce structural universals to non-structural ones. I do not even think that their analysis of the structural universal displayed by all methane molecules is adequate. A more adequate analysis would have to describe the structure as tetrahedral, with the four hydrogen atoms at the apices and the carbon atom centrally placed within the tetrahedron. Bigelow and Pargetter's account does not imply that methane molecules have this structure. They might indeed have any of a great many structures compatible with the CH_4 molecular formula and the stipulated bonding relationships. If the wheels of a car were hydrogen atoms, and its chassis were a carbon atom, then it would have the structure of a methane molecule, if the wheels were stuck on appropriately.

The methane structure is in fact a species of a generic kind of tetrahedral molecular structure other species of which include the structures of carbon tetrachloride (CCl_4), monosilane (SiH_4), carbon and silicon tetrafluoride (CF_4 and SiF_4), and stannic chloride ($SnCl_4$). All of these kinds of molecules are tetrahedral, with a central quadrivalent atom that is bonded to monovalent atoms at the apices. Any adequate ontology should thus be able to account for a range of generic structural universals, each having a variety of infimic species. And I see no reason to think that such universals can be ontologically reduced to non-structural property universals of the more traditional kind.

I also propose that we recognize the existence of a basic hierarchy of natural kinds of objects, wherein individuals are distinguished by their essential properties and structures, and are related to each other by species and other family relationships. The members of this hierarchy of natural kinds are more or less specific kinds of objects or substances, the instances of which are just the objects or substances that belong to these kinds. The most specific (or infimic) species, in the hierarchy of natural kinds of objects are the simplest kinds of substances. Plausibly, things such as electrons and protons belong to this class, although there may well be more fundamental things than these. The instances of each infimic species of natural kind in the category of substances must all be essentially the same. For if they were not, then the species would have sub-species. They are therefore like classical universals (which have no sub-species), although, as we shall see, they are not property universals either. The more general species in the hierarchy of natural kinds of objects or substances, such as atoms and molecules, are generic kinds. These are formally like generic universals, or determinables, although the category to which they belong is not that of properties and relations, but that of substances. Accordingly, we may refer to the members of the hierarchy of objects or substances as more or less general "substantive universals."

The hierarchy of natural kinds of objects or substances is not evidently assimilable to the hierarchies of properties and relations that exist in the world. For the members of these hierarchies have different kinds of instances. The instances of property and relation kinds are tropes; those of substantive universals are objects or substances.

A hierarchy of natural kinds of tropes is needed to account for the kinds of properties and relations that exist in the world. The infimic species of this hierarchy are property and relation universals that have identical instances. The tropes of any given determinate value of a natural quantity, for example, are all identical. So every determinate value of every natural quantity is an ultimately specific kind of property or relation. It is a determinate universal. But the quantities themselves are generic universals, since the relationship between a quantity, such as *having mass*, and a specific measure of that quantity, such as *having a mass of two grams*, is not that of a universal to an instance of that universal, as the theory that quantities are higher-order universals requires. Rather, the relationship between a quantity and a specific quantitive property of that kind is one of genus to species. This should be clear from the fact that all instances of having a mass of two grams are also instances of having mass.

The instances of any specific quantitative property are *property instances;* they are not the objects in which these instantiations may occur, because the objects in which such a specific universal is instantiated need not all be the same. Two grams of water is not identical to two grams of alcohol. They may indeed be identical only in mass. It is only the mass of the water and the mass of the alcohol that must be identical if both are exactly two grams in mass – nothing else. Therefore the instantiations of the classical universal in these two cases are not the samples, but the masses of the samples.

I argue that objective quantities such as rest mass should be thought of as natural kinds of tropes, and specific quantitative properties, such as that of having a rest mass of two grams, as the infimic species of such natural kinds. The case for thinking of such quantities and quantitative properties in this way goes like this. These quantities satisfy all of the requirements on natural kinds.

1. They are *objective:* Whether an object has a rest mass of two grams or not is not a matter that depends in any way on us.
2. They are *categorically distinct:* The rest mass of any object is categorically distinct from its charge and from every other quantitative property.
3. The difference between the mass of an object and its charge is an *intrinsic difference,* not an extrinsic one.
4. Tropes of the universals "having a rest mass of two grams" and "having a rest mass of four grams" are both species of the generic universal "having a rest mass," so the *speciation* requirement is satisfied.
5. Rest mass and charge are both species of causal powers, as opposed to shape, which is not. All instances of rest mass and of charge are therefore instances of causal powers. The *hierarchy* requirement is thus satisfied.
6. Rest mass and charge both have *real essences* – namely, in the dispositions they underpin – as will be argued in Chapter 3.

It is necessary also to recognize the existence of a hierarchy of natural kinds of events and processes. I am not sure that this hierarchy is independent of the hierarchies of objects and properties. It may be that every natural kind of event or process is necessarily a display of some natural kind of dispositional property, or set of dispositional properties in natural kinds of circumstances. If so, then the existence of a hierarchy of natural kinds of events or processes will be a simple corollary of the existence of a natural hierarchy of causal powers and other dispositional properties. But the required category of natural kinds of circumstances presents a problem for this view. First, there might be some natural kinds of events or processes that are not just kinds of displays of natural dispositional properties in natural kinds of circumstances. The Big Bang, for example, might be an event of a natural kind that cannot be characterized in this way. Second, Swinburne's regress argument (Chapter 3, Section 3.12) might well present a problem for this reducibility thesis, since an ontology that includes natural kinds of circumstances would also seem to require kinds of events that are not just the displays of dispositional properties. But let me skirt around the issue, and assume for now that the hierarchy of natural kinds of events and processes is independent of the other two hierarchies. It is obviously not wholly independent of them, and it may not be independent at all. The fact that there is such a hierarchy seems clear enough.

The different chemical reactions, for example, are all natural kinds of processes. Thousands of them are described in chemistry textbooks. These processes, like the chemical elements and compounds involved in them, obviously satisfy all of the requirements on natural kinds. They are objectively occurring, categorically and intrinsically distinct from each other, belong in natural hierarchies of greater or less generality, and so bear species and other family relationships to each other, and they have real essences, most of which are now fairly well known. When hydrogen combines with chlorine to form hydrogen chloride, we can describe the essential nature of this process, and thus explain why the process occurs when it does, and in the way that it does.

An ontology of localized states of affairs, which envisages a world of logically independent atomic objects instantiating infimic species of various quantitative and qualitative universals, does not seem to be rich enough to do justice to the hierarchical structure of reality. It lacks natural kinds, and has no obvious place for the species relationship. Hence it seems not to be adequate to account for the complex patterns of necessary connections involving this relationship that we find in nature. If we suppose the world to consist just of specific properties and relations and the particulars in which they are instantiated then we cannot readily explain the existence of hierarchies of natural kinds of objects or of generic properties and relations. Their existence would appear to be just a cosmic accident. Nor can we explain the existence of hierarchies of natural kinds of processes – entities that would appear to have no place at all in the traditional ontology, but that nevertheless are the subject-matter of most of the laws of nature.

A much richer ontology than any neo-Humean one is required if we are to account adequately for those broad features of reality which are summarized in the Basic Structural Hypothesis. I do not say that the rich ontology of natural kinds, which I have outlined here, with it three kinds of generic universals, is strictly necessary for the purpose. But I do claim that it is sufficient, and should therefore be a starting point for ontological inquiry. It would certainly be of interest to be able to see more clearly how substantive, dynamic, and property universals are related to each other, and to discover whether any further ontological reductions, either in or between the different categories of universals, can be achieved. We might in the end even be able to get down to an ontology something like Armstrong's (although I very much doubt it). But we should not start from such a position, at least not if we wish to explain the sorts of broad features of reality that science has laid bare. The pattern of relations between natural kinds, their species, and their instances is given in Table 1.

At the summit of each hierarchy there is a global kind. These kinds have no accepted names, because there are no other kinds of objects, events, or properties known to us with which to contrast them. The global kind, in the hierarchy of objects or substances, is a natural kind which is postulated to include among its species all other natural kinds of objects or substances existing in the world. The global kind in the category of events and processes is a natural kind, which is supposed to include all other natural kinds of events or processes occurring in the world. The global kind in the category of properties or structures is the supposed natural kind, which is supposed to include all other natural kinds of properties, relations, and structures existing in the world. In other worlds, there are presumably objects, events, properties, and structures that do not belong to any of these global kinds. In Chapter 7, I will argue that the most general of the laws of nature are necessary propositions describing the essential natures of these global kinds.

For now, let us focus on the more specific natural kinds. In the theory being developed here, substantive kinds, dynamic kinds, and property kinds necessarily have different sorts of instances. Substantive natural kinds are always instantiated by objects or substances,[2] dynamic natural kinds are always instantiated by events or processes, and properties and relations are always instantiated by their tropes.

Substantive universals are not instantiated *in* things in any sense other than a locational one. Instances of the natural kind, *hydrogen atom,* for example, are to be found in hydrogen compounds or in gases containing hydrogen. But the generic universal *hydrogen atom* is not instantiated *by* hydrogen compounds or gases containing hydrogen. Nor is it instantiated *in* a single hydrogen atom. Rather, it is instantiated by the hydrogen atom itself. To suppose that the substantive natural kind *hydrogen atom* is instantiated *in* a single hydrogen atom is to treat the substantive universal, *hydrogen atom,* as if it were a property universal – that is, *being a hydrogen atom.* But what could the substance be in which the trope of being a hydrogen atom is located? Is it some neutral kind of substance that just so happens to have the property of being a hydrogen atom? Or is it a substance that has this supposed property necessarily? It might be the latter,

Table 1 The Natural Kinds

Levels of generality	Substantive natural kinds	Dynamic natural kinds	Natural property kinds
Universals of very general kinds	Elements Compounds Fundamental particles	Causal interactions Energy transfer processes Causal processes	Dispositional properties Categorical properties Spatial and temporal relations
Universals of more specific kinds	Inert gases Sodium salts Baryons Leptons	Chemical reactions Ionizations Electromagnetic radiations Diffractions	Mass Charge Field strength Shape Spatiotemporal interval
Infimic species of the more general kinds	Helium atoms Sodium chloride molecules Neutrons Electrons	$H_2 + Cl_2 \Rightarrow 2HCl$ $H_2 \Rightarrow 2H^+ + 2e^-$ Photon emission at $\lambda = 5461\text{Å}$ from an atom of Hg Diffraction of light with $\lambda = 5461\text{Å}$ at the surface of a quartz crystal, where the angle of incidence is $5°$ to the normal	Unit mass Charge of 2e Unit field strength Spherical shape (Note: The infimic species in this category are the kinds of properties called "classical universals" in the text)

The instances of the natural kinds

Things that instantiate the kinds in the above categories	Objects or substances that belong to the infimic species of things in this category	Events or processes that belong to the infimic species of things in this category	Tropes of the infimic species of the natural properties of things (Note: There are no freestanding tropes)

if the supposed property of being a hydrogen atom were just a conjunction of causal powers. For then one might think of the property of being a hydrogen atom as a simple conjunctive universal (which is the simplest kind of structural universal). A thesis of this kind is certainly plausible for the fundamental constituents of matter, which presumably have no structure. But hydrogen atoms and the atoms and molecules of more complex substances all have structures, some of them quite complex. Yet their structures are no less essential to their identities as kinds. Therefore, to maintain the thesis that there is a property of being a hydrogen atom, and not just a natural kind consisting of hydrogen atoms, one would need a decent reductive theory of structural universals of the kind sought by Bigelow and Pargetter – one that would allow us to think of structural universals as first-order properties like mass or charge, but just a lot more complex. However, no such theory is yet in sight.

It is also a mistake to think of dynamic universals as properties of, or relations between, events. Events of electron-positron annihilation, for example, are members of a natural kind. Therefore, there is a dynamic universal of *electron-positron annihilation*. But there is no *property* of being such an event, because there is no object or event in which a trope of such a property could inhere. Properties have to be instantiated *by* their tropes in the things that have them. But in this case, there is nothing for a trope of this dynamic property to inhere in. Likewise, we may ask, what is the radiation that happens to have the property of being electromagnetic radiation? Is it radiation which might have been, but happens not to be, some other kind of radiation? If so, then what can we say about this neutral sort of radiation independently of the fact that it happens to be electromagnetic?

For these reasons, and for others that will be elaborated (in Section 2.8), it would appear to be wrong to try to assimilate substantive and dynamic kinds to property kinds. The natural kind *hydrogen atom* cannot possibly be instantiated by anything other than a hydrogen atom, and the kind *electromagnetic radiation* cannot possibly be instantiated by anything other than electromagnetic radiation. And the alleged properties of being a hydrogen atom and of being electromagnetic radiation cannot plausibly be said to be instantiated *in* anything.[3]

It follows that substantive and dynamic natural kinds are not just conjunctive universals, the conjuncts of which are properties or relations. Conjunctive universals would have the wrong sorts of instances. A conjunctive universal, if there were such a thing, would have to be instantiated *in* something, not by an object or a process. Moreover, all natural kinds, other than the very simplest ones, have instances that are structures of instances of other natural kinds. Consequently, natural kinds cannot be constituted by logical or mereological operations on ordinary universals. They cannot be constituted in this way because universals are not spatially locatable, and they cannot be duplicated. So they cannot be put together in the sort of way that is required to construct compound universals, the instances of which consist, say, of two things of one kind and one of another arranged in a certain way. A water molecule, for example, has hydrogen and oxygen nuclei as components. But the natural kind *water* cannot be said to have the supposed universals *being a hydrogen nucleus* and *being an oxygen nucleus* contained within it in the ratio of two to one. Of course, if the supposed compound universal *being a water molecule* and others like it could be constituted somehow, then natural properties might be ontologically more fundamental than natural kinds. But it seems that they cannot. Hence, an ontology that includes natural kinds, and is not reducible to an ontology of universals and the particulars in which they are instantiated, would appear to be needed to explain the structure of the world.

I do not wish to argue here that the ontology of natural kinds cannot be reduced in any way to a simpler ontology. Keith Campbell (1990) would argue that tropes are ultimately enough for ontology, given a few primitive relationships of exact resemblance, compresence, and perhaps one or two others. Tropes have the advantage of being spatially located, and relatable by the primitive relationships. Caroline Lierse is sympathetic to Campbell's project, because it seems to her to be the only possible "bottom up" ontology in which there is just a single class of ultimately simple basic kinds of entities. However, Campbell would throw the whole weight of explaining the structure of reality onto the primitive relationships and set-theoretical or similar operations he requires. On Campbell's theory, for example, a continuous quantity comes out as an infinite disjunction of equivalence classes of exactly similar tropes (one for each value of the quantity). Whether such concepts can carry the weight they are required to bear is doubtful because the universals the theory is seeking to reduce to tropes, and primitive relations between tropes, have a fundamental theoretical role that this theory does not reflect. And it does nothing to explain the broad facts about the world alluded to in the Basic Structural Hypothesis.

2.4 ACCIDENTAL, INCIDENTAL, AND ESSENTIAL PROPERTIES

We can say quite generally that if two things belong to the same natural kind, then there must be certain respects in which they are intrinsically the same. For what makes something a natural kind cannot have anything to do with the accidents of its location, surroundings, or history. Moreover, the class of respects in which they are intrinsically the same must be sufficient to distinguish things of that kind from things of any other kind. Atoms of uranium, for example, are intrinsically the same in ways that distinguish them categorically from all other kinds of atoms – for example, they all have atomic number 92.[4] Hence, atoms of uranium belong to a natural kind. However, not all atoms of uranium are intrinsically identical. For two atoms of uranium may differ in atomic weight. Moreover, the intrinsic differences between them may make a very big difference in the way the two atoms are disposed to behave. U^{235} is unstable, and may undergo fission; U^{238} is relatively stable, and is disposed to decay in other ways. The natural kind, *uranium,* therefore has at least two clearly distinct species. (Actually it has more than two.)

We call those natural kinds which have distinct species or subspecies *generic kinds.*[5] Then the members of any generic kind K must have certain common intrinsic properties or structures P that make them the kinds of things they are. These common properties or structures will together constitute the real essence of the genus K. Let S be any species of K. Then the members of S must not only have the generic essence P, they must also have certain other intrinsic properties or structures Q that distinguish the members of S from the members of all other species of K. The set of properties P+Q will then be the real essence of the species S, and the properties Q will be the essential properties that are distinctive of the members of S in K.

Traditionally, essential properties are contrasted simply with accidental properties. Therefore, relative to the generic kind K, the properties Q that are not essential must be classed as accidental. For example, this crude classification of properties as essential or accidental would require us to say that an atom of U^{238}, qua atom of uranium, has atomic weight 238 only accidentally, for it could be transformed into an atom of U^{234} by emission of an α-particle, followed by subsequent emissions of two β-particles. Atomic weight 238 is therefore a property that an atom of uranium could have or lack, or gain or lose, without ceasing to be an atom of uranium. Arguably, it could gain or lose this atomic weight without ceasing to be the atom of uranium it is. Therefore, by all of the usual criteria for distinguishing essential from accidental properties, atomic weight must be classed as an accidental property of the atom, qua atom of uranium.[6] However, U^{238} is a species of uranium that is itself a natural kind, and vis-à-vis this kind, atomic weight 238 is an essential property. For an atom of this isotope must have this atomic weight, and it could not lose this property without ceasing to be an atom of this isotope, and hence ceasing to be a thing of the kind it is. Therefore the traditional distinction requires us to say that while the property of having atomic weight 238 is an essential property of an atom of U^{238}, it is only an accidental property of an atom of uranium.

I think that this way of speaking is confusing and glosses over an important distinction. The distinction glossed over is between properties that are essential at some level of specificity of natural kind and properties that are not essential at any level of specificity. The property of being an atom in this rod, for example, is not an essential property of any natural kind at any level of specificity. Of course, we can always define a kind for which being an atom in this rod is part of its *nominal* essence. For example, the class of atoms in this rod has the property of being an atom in this rod as its nominal essence. But the class of atoms in this rod is not a natural kind since membership of this class does not depend on the intrinsic properties or structures of the atoms in question. It depends only on where they are, and location is an extrinsic property, not an intrinsic one.

To overcome this difficulty, I propose a three-way distinction. If a property Q is not essential to a natural kind K, but is essential to a natural species of K, then I propose to say of any member of K that has Q that it is a member of K that has Q incidentally, and is therefore a member of a natural species of K which has Q essentially. The following two statements are therefore logically equivalent:

1. x is a member of K that has Q *incidentally*.
2. x is a member of a natural species of K that has Q *essentially*.

Both must, however, be clearly distinguished from:

3. x is a member of K that has Q *accidentally*.

For to say that x is a K that has the property Q accidentally is to imply that there is no species of K that has Q essentially. For example, I shall say of a given uranium atom that is in fact located in a particular rod that it has the property of being so located accidentally. However, if it also happens to be an atom of U^{238}, then that is an incidental fact about this particular uranium atom. Other uranium atoms in the rod will have different atomic weights. Some will incidentally be atoms of U^{239}, and others, perhaps, atoms of U^{234}. What is essential to anything's being a uranium atom of any kind is its having atomic number 92 and the characteristic electron structure of all such atoms.

Without a three-way distinction like this, one cannot distinguish adequately between the accidental properties of the more inclusive natural kinds and the distinctive intrinsic properties of their natural species. But this distinction is obviously an important one, because these distinctive intrinsic properties are among the essential properties of these species. Therefore, where there are hierarchies of natural kinds, this extra class of properties must be recognized. As we shall see, this extra category is also needed for the description of spectral kinds.

It is evident that the natural kinds of the physical sciences frequently permit continuous ranges of intrinsic variation among their instances. This is true of spatial and temporal intervals, and it appears to be the case for some dispositional properties, such as field strength. Hence, most natural kinds of physical processes will also permit continuous ranges of intrinsic variation. I call natural kinds like these *spectral kinds*. Where we are dealing with a spectral kind, there cannot be any species other than the absolutely specific ones – that is, species defined by completely precise permitted values of the variables. Hence the universals existing at the limits of specificity of spectral kinds will mostly be of theoretical interest only since they will rarely, if ever, be instantiated. Moreover, any division of the continuous range of permitted intrinsic variation that is any more coarse-grained than the limiting one would have to be either arbitrary or based on gaps accidentally occurring in the range. In neither case would the division of the kind be a division into natural species.

The variable quantities whose precise values distinguish the various limiting species within the spectral kinds are not just accidental properties of these kinds, since they are the intrinsic properties that locate the various instances within the spectrum of possibilities. They are the properties that make the instance precisely the kind of instance it is. The precise values of the variables in a spectral kind are therefore incidental to the kind, but essential to its being precisely the species of the kind that it is. To illustrate: Electromagnetic waves all have the essential property of being propagated according to Maxwell's equations. All electromagnetic waves must be so propagated, and nothing other than an electromagnetic wave can be propagated in this way. But electromagnetic waves generally differ from each other in shape, frequency, and amplitude. Clearly, any particular electromagnetic wave must have particular values for these variables. It must have a certain frequency and amplitude, and be of a certain shape (spherical, planar, and so on). These additional properties, which are distinctive of the particular kind of wave we are dealing with, and which are essential to its being precisely the kind of wave it is, are incidental, rather than accidental, properties of the electromagnetic wave. Continuous quantities, like field strength and spatio-temporal interval are natural kinds that have precise field strengths and spatio-temporal intervals as their ultimate species. Hence, these quantities must be classed as spectral kinds. The instances of these spectral kinds are the tropes of these quantities occurring in the regions of space-time or in the pairs of events that possess them.

To simplify, let us assume that mass (intrinsic or rest mass) is a continuous quantity. Whether it is or not, I do not know. But for the purposes of discussion this does not matter. Let G be the generic universal *mass* and S the specific *mass m*. Then S is a species of G. Let A be any object which, as we say, has mass *m*. Then A is an object in which a trope of S is instantiated, and since S is a species of G, A must also be an object in which a trope of G is instantiated. It should be noted that S is not an instance of G; it is a species of G. It should also be noted that A is not an instance of either S or G. It is just an object in which a trope of S, and hence of G, is instantiated.

The difference between an instance of a kind and a species of that kind is important, and must be stressed. An instance of a given kind is something in the category of things over which the kind ranges. Thus, an instance of a natural kind of animal must be an animal. It cannot be a number or a tree. An instance of a natural kind of process must be a process. It cannot be an object or a property. A species, on the other hand, must itself be a kind, and it must be one that ranges over the same category of things as the kind of which it is a species. I take the view that the quantity *mass* has many different species. Some of these, like *electron mass* and *proton mass,* are natural, ultimately specific, species of mass. Hence, they are classical universals. Others, like *mass of more than four grams,* though a species of mass, is not a natural species. Instances of *electron mass* are to be found in all electrons and positrons, instances of *proton mass* in all protons. But note that the instances are not the electrons or the protons themselves: They cannot be, for these are the instances of very different kinds of things — namely, *fundamental particles.*

In Aristotelian theory, which is the starting point for most philosophical debate about natural kinds, the instances of natural kinds are generally considered to be ontologically more fundamental than the kinds they instantiate. They are more fundamental in this sense: The kinds would not exist if no instances of them existed. Most philosophers would also argue that natural species are ontologically more fundamental than the natural families to which they belong. That is, the families would not exist if no species belonging to them existed. It is arguable whether one instance is sufficient to establish the existence of a kind, or one species sufficient to establish the existence of a family. However, this issue is not one that need concern us here. What is common to Aristotelian theories about natural kinds is this: The existence of a natural kind depends on the existence of at least one instance of it, and the existence of a natural family, or genus, depends on the existence of at least one species in this family.

In Plato's theory, which is the main alternative to Aristotle's, the kinds, or "Forms," as Plato calls them, have ontological priority. They are eternal, and would exist whether or not anything in the ordinary world of experience participated in them. The material objects that belong to the actual world cannot instantiate the Forms, as Aristotle's universals might be instantiated, but they might more or less closely resemble them, and to the extent that they do, they are said to participate in them.

The main argument in favor of an Aristotelian theory is an epistemic one. The instances of a property or kind are there to be seen or felt or measured, whereas the property or kind itself can only be studied by studying its instances. Hence an uninstantiated property or kind cannot be studied or investigated at all. And if it cannot be studied or investigated, then it is hard to see why we should believe in it. An uninstantiated property or kind would seem to be more like a human construction or invention than anything that really exists.

A second argument in favor of an Aristotelian theory is one from commonsense realism. Anyone who is armed with what Russell would have called "a robust sense of reality" would certainly suppose that what is concrete and known to exist by observation or experiment is ontologically more fundamental than what is abstract and incapable of being known in these direct ways. Aristotle's assumptions about ontological priority are consistent with this commonsense position; Plato's are not.

Nevertheless, the arguments about ontological dependence are not all favorable to an Aristotelian position. *First,* there are reasons to believe that there are kinds that might never be instantiated. For example, there could well be a transuranic element, capable of existing for some tiny fraction of a second, that has never existed, and never will exist, either here on earth or anywhere else. And just as Mendeleyev, in constructing his periodic table, was able to make predictions about elements yet to be discovered on earth, today's scientists might be able to say what the properties of this transuranic element must be. If so, then it will be the case that although no atoms of this element have ever existed, and none ever will, some of their essential properties are known. Second—and this is crucial—it would seem to be a genuine scientific question whether the atoms of this transuranic element really would have these properties—that is, whether the scientists are right. But how could we be right or wrong about the essential properties of a non-existent kind of thing?

This argument, although suggestive that natural kinds might exist uninstantiated, is inconclusive, and I do not wish to put much weight on it. I would not wish to open the door to an unlimited range or variety of uninstantiated and unknowable universals. One would at least have to have some principled way of deciding which uninstantiated universals exist and which do not. That is, one would have to have a sparse theory of uninstantiated universals if any such theory were to be acceptable. I know of no such theory. The Aristotelian position has the advantage on this point, and I see no sufficient reason to abandon it. It is important to stress, however, that the doctrine of *universalia in rem* does not require that properties or natural kinds be ontologically dependent on their instances. For while it might be agreed that a natural kind or property could not exist uninstantiated, it would also have to be agreed that no instance of any given property or kind could exist if the property or kind of which it is an instance did not exist. The property or kind would have to exist even if the property or kind instance was quite unique.

2.6 THE THEORY OF ONTOLOGICAL REDUCTION

We are all familiar with the idea that the existence of some things depends on the existence of others. The existence of living things, for example, depends on the existence of organic molecules. Organic molecules, in their turn, depend for their existence on carbon atoms. Where one kind of thing thus depends for its existence on another kind of thing, we speak of *ontological dependence,* and the process of tracing lines of ontological dependence down to ontologically more fundamental kinds of things as that of *ontological reduction.* The question we wish to tackle here is: What concepts of ontological dependence and reduction do we need for natural properties and presumably for natural kinds and relations?

To say that a property P_1 is ontologically dependent on a property P_2 is at least to say that the property P_1 would not exist if the property P_2 did not exist. Therefore, we should expect the relation of ontological dependence for properties to be closely related to that of supervenience. However, supervenience is not quite the relation we need. For, as supervenience is commonly defined, the subvenient base for the property P_1 on which P_1 may be supposed to be ontologically dependent may be much too broad. It may, for example, be the class of all physical properties, and so contain many properties that are simply irrelevant to whether anything has the property P_1. We propose the following: A class of properties B is a *relevant subvenient basis*[7] (RS-basis) for a property P if and only if something's having all of the properties in B is a sufficient, but not a redundantly sufficient, condition for its having the property P. Conversely, we say that a property P is *ontologically dependent* on the properties B if and only if the properties B belong to every RS-basis for P.[8]

Given these definitions, it is evidently possible for a property P to have one or more RS-bases. If a property P must always have the same RS-basis B, – that is, if nothing could have the property P without having all of the properties in B, then P may be said to be *ontologically reducible* to B. On the other hand, if it is possible for different occurrences of a property P to have different RS-bases, then P is not ontologically reducible to any of its RS-bases. No such reduction is possible because the various occurrences of P are not necessarily manifestations of the same basic state of affairs. What unites the various occurrences of P, and gives rise to their common classification, can only be some more or less superficial similarity of appearance, behavior, or function. It cannot be the case that all of the things having this property are essentially the same.

Also given these definitions, it is possible that there are properties that have no RS-bases. All such properties may be said to be *ontologically basic properties*. Plausibly, such ontologically basic properties must exist, although we may not yet know what they are. But if any property P is not an ontologically basic property, then its instances must be ontologically dependent on properties that are subvenient to P and which therefore could be instantiated independently of P.

What then should we say is the relation of ontological dependence between a quantity, say mass, and all of the existing species of this quantity? For reasons given, we adopt the Aristotelian theory of the existence of properties – that is, the doctrine of *universalia in rebus*. Then it is not true that all possible species of the spectral kind mass exist (or ever have or ever will exist). To refer back to an earlier example, it is almsot certain that there is no object anywhere in the universe with a rest mass of $0.5\ m$, where m is the mass of the electron. Therefore, this quantitative property does not exist. But mass exists in this world, and plausibly it could exist in many different worlds in which the mass spectrum has other gaps. Therefore, the set of all species of mass existing in this world does not constitute a unique RS-basis for mass. On the contrary, it is at least plausible to suppose that mass could exist even if none of the existing species of mass existed. Therefore the quantity mass cannot plausibly be supposed to be ontologically dependent on any of its species. On the contrary, since mass must exist in every world in which any species of mass exists, the species must depend on the quantity.[9]

The best way of avoiding this conclusion, if one wishes to, is to follow Plato, and insist that the existence of a quantity (or any other spectral kind) entails the existence of all of its species. For then we can say quite generally that spectral kinds and their species are equally fundamental. There is, however, no plausible account on which a given natural quantity (or any other spectral kind) turns out to be ontologically dependent on its natural species. This is true even if one adopts a Platonist thesis about universals. Hence, "more general" does not imply "less fundamental" in ontology.

The conclusion that quantities and other spectral kinds are no less fundamental than their species is initially surprising but also pleasing. It is surprising, because it runs counter to our atomistic intuitions about ontological dependence. In biology, it is plausible to suppose that kinds are just sets of species whose members all happen to share some common properties. But if the members of natural kinds have essential properties or structures, qua members of those kinds (as I am supposing), then the existence of just one member of a natural kind entails the existence of all of the more general kinds of which it is also a member. For the properties and structures that would make it a member of that kind must include those that would make it a member of any of these more general kinds. It is pleasing because it goes some way toward explaining why quantities and other natural spectral kinds are so important in science. If quantities are at least as fundamental in ontology as any of the specific quantitative properties that are instantiated in the world, and natural kinds of processes are at least as fundamental as any specific cases of such processes, then natural science is bound to be concerned with them and with the relations that hold between them. No other ontology that I know of explains this focus of science as well.

2.7 REAL PROPERTIES AND PROPERTY IDENTITY

Often, things that are intrinsically different are said to be the same. They are classified together because they affect us similarly (for example, the use of the same color terms to describe transmitted, reflective, and radiant colors) or serve some common purpose (for example, in highlighting or illuminating). On the other hand, things are often distinguished from each other in ways that do not correspond to any real distinctions in nature (for example, the distinctions between adjacent colors or species).[10] The distinctions are made according to how we perceive things or what we take to be distinguishing characteristics, although ontologically the distinction we make may be arbitrary.

To explain similarities or differences that depend in some way on the sorts of beings we are – for example, on how we perceive the world, on what our interests are, on the uses we have for things, on the language we speak, and so on – there is no need for us to believe in natural or objective properties. For the classifications and distinctions we make that depend on such factors as these do not necessarily correspond to any intrinsic similarities or differences between the things that are classified together or distinguished.

Some philosophers and many sociologists would no doubt deny that there are any natural or objective properties or relations. For example, many would argue that since distinctions are never forced on us, we are free to recognize any classes of things we wish, depending on what our interests happen to be. Therefore, the classification system we have learned to use and are still developing has no basis in any reality that does not include ourselves as classifiers. Therefore, there is no objective system of classification of things for a theory of objective properties and relations to explain. The only question that needs to be answered, it is sometimes said, is the sociological one: How did we come to construct the classification system that is now embedded in our language?

Needless to say, I reject this view. I reject it for the reason that scientific realists reject constructivist theories about theoretical entities. Realism about the properties and relations postulated by scientists to explain causal and stochastic processes provides the best explanations of these processes. The alternative of supposing that the postulated entities, with their postulated properties and relations, are not objective features of the world, but constructs that we find more or less useful for systematizing our experience, only creates a new problem – namely, that of explaining the appearance of objectivity. For how can this be explained unless it is supposed that human beings have a common nature and that we are reacting to the same world? Moreover, the supposition that we have a common nature is precisely the supposition that we ourselves are the bearers of objective properties. Therefore if we deny that there are any objective properties in nature, we can only explain the contrary appearance by supposing that there are objective properties in nature after all.

The main case for realism about the kinds of properties and relations that are postulated to explain regular patterns of behavior is similar to that for realism about the more substantive kinds of theoretical entities that occasionally feature in such explanations. For example, we should be realists about atoms and molecules and electromagnetic fields, because the world certainly behaves as if things of these kinds existed. Likewise with the supposedly objective properties of things. A piece of sealing wax that has been

85

rubbed with fur is disposed to behave as if it were negatively charged, and the best explanation of this fact is simply that it really is negatively charged. The behavior of the sealing wax could in principle be explained in some other way, for example, by the action of a deceptive god. But there is no serious alternative to the "negative charge" explanation. Therefore the case for realism about such a property is similar to that for realism about such substantive theoretical entities as atoms, molecules, and electromagnetic fields, and it is no less powerful.

Natural properties and *natural relations* are postulated to explain the intrinsic similarities and differences between things in nature, independently of how we may conceptualize nature. If two things have the same mass, then they are intrinsically the same in this respect. If two pairs of events are separated by the same spatio-temporal interval, then there is an intrinsic similarity between these two pairs of events. Such properties and relations exist in nature independently of us, and independently of our perceptions, customs, and interests.

Ontologically, the most fundamental entities in the category of properties and relations are not the specific properties and relations that things may have, or bear to each other, but the quantities of which these specific properties and relations are infimic species. Included among these, we may suppose, would be those quantities that are conserved according to the conservation laws, and those kinds of relationships of a spatio-temporal or numerical character that things or groups of things may bear to each other. I postulate that the spatio-temporal and numerical kinds of relations are irreducibly relational – that is, not ontologically dependent or supervenient on non-relational kinds or properties.

The identity of a property is not determined by the set of things that have it. For there is no reason why two properties should not happen to have the same extension. To use the standard example, every chordate (animal with a heart) might happen to be a renate (an animal with kidneys), and conversely. But the property of having a heart is distinct from that of having kidneys. The properties are distinct because (it seems) there could be a world in which there were chordates that are not renates, or renates that are not chordates.[11]

To deal with this problem, it is often said that the identity of a property must depend, not simply on the set of things that possess it in the actual world, but on the set of all things, in all possible worlds, that possess it. That is, the criterion for property identity is often supposed to be necessary co-extensionality.

If "necessary" here means "logically necessary," then this criterion for

86

property identity might well be formally correct, although there are cases that suggest this is not so. For example, the property of being a three-sided polygon is not obviously the same as that of being a polygon with exactly three vertices, even though these "properties" are necessarily co-extensional. However, defenders of the necessary co-extensionality criterion need not be swayed by this example, and I am inclined to think that they should not be swayed. For one may well have a theory of property identity according to which there is only one property (or more strictly speaking "kind of property") here – triangularity, and argue that this is a property which manifests itself in more than one way. So the question of whether necessary co-extensionality is formally correct as a criterion for property identity remains open.

The question we have to ask is: What makes one property the same or different from another? And this question would remain even if necessary co-extensionality were agreed to be a necessary and sufficient condition for property identity. If this condition were formally correct, why should it be? Why may different properties not have the same extensions in all possible worlds?

If the criterion for property identity is supposed to be necessary co-extensionality in some weaker sense of "necessary" – for example, coextensionality in all physically possible worlds – then the criterion would appear to be unsatisfactory on almost any theory of property identity. For there are manifestly distinct properties that are evidently co-extensive in all physically possible worlds. Atoms of hydrogen, for example, are the same as each other in a number of different respects. Moreover, there are several respects in which hydrogen atoms differ from all other things. For example, they have their own unique nuclear structure, their own unique electron structure, and their own unique optical spectrum. Moreover, it is physically impossible for any atom that has any one of these properties to lack either of the others. An atom cannot have the electron structure of hydrogen without having hydrogen's nuclear structure, or the capacity to produce the characteristic optical spectrum of hydrogen. Equally, it is physically impossible for any atom that has the nuclear structure of hydrogen to lack hydrogen's electron structure, or to give rise to a different optical spectrum.

Even if a materially adequate definition of property identity could be given in terms of necessary co-extensionality, the point has to be stressed that such a definition would be metaphysically unsatisfactory. For the extension of a property in any given world is just a contingent fact about that world. Its extension is not of the essence of the property. To think that the identity of a property depends on its extension in some, or even in all possible worlds, is to think of properties as things like sets. But properties are not sets, and cannot be satisfactorily reduced to sets.[12] A set is defined by its extension, so new members cannot be added to or taken away from a set without changing its identity. The membership of a set is of its essence. But properties are not like this. The identity of a property does not depend on what things possess it, whether in this or in any other possible world, and things can gain or lose properties without these properties in any way changing their identities.

The identity of a set depends on its membership, and the proper subsets and super-sets of any given set are different sets. Hence, a set exists if and only if all of its members exist. But the identity of a property does not depend on what possesses it. If you are an Aristotelian, then you will say that a property exists provided there is at least one thing that actually has it; it does not matter what that thing might happen to be. If you are a Platonist, then you will allow that a property might exist, although nothing actually has it. But neither Aristotelians nor Platonists think that the identity of a property is tied to its extension. The set of bearers of a given property is just the contingent extension of the property. The property could have had some other extension.

The identity of a property thus does not depend on its extension in this, or in any other, possible world; it depends, rather, on what it is, in the case of a categorical property, and on what it *does,* in the case of a causal power, capacity, or propensity.[33] In neither case does its extension determine its identity. For example, the distinction between gravitational mass and inertial mass is not extensional. The inertial mass of a body is its capacity to resist acceleration; its gravitational mass is its power to attract other bodies gravitationally. The two may be the same, even necessarily the same, but they are still quite distinct properties. I say this because, at least in the cases of causal powers, capacities, and propensities, property identity depends on sameness of causal role. But gravitational mass and inertial mass clearly have distinct causal roles. Even in the world of General Relativity, there is a quantity that we may call "relativistic gravitational mass" – which is a body's power to curve space-time – that is quite distinct from its inertial mass, which is still just its capacity to resist acceleration.[14]

Extensional semantics hold such sway in philosophy these days that many philosophers not only think of properties as sets of individuals, they think of the instances of properties as the individuals that belong to these sets. This is wrong on both counts. Not only are properties not sets, as we have seen, but the instances of properties are not the individuals that have them. Properties are characteristics that are instantiated in individuals, not by them. The instances of properties are *property instances*. They are what Donald Williams and Keith Campbell call "tropes." The identities of the tropes instantiating a given property are normally supposed to be tied to the individuals that have this property. That is, if X has the property Q, then the trope of Q that is instantiated in X is the Q-ness of X. Hence, nothing other than X could have this particular trope. But the tropes are not the individuals in which the properties are instantiated. If the electrons A and B are two different individuals, then the mass of A and the mass of B are two instances of the same infimic species of the quantitative property mass – namely, unit mass. But A and B are not themselves instances of this property.

It is helpful to think of the tropes of the natural properties possessed by any object as the set of causal powers and structures located in the object. They are not these causal powers and structures considered abstractly, but the concrete instances of the causal powers and structures occurring in the object.

2.8 PROPERTIES AND PREDICATES

There is an important distinction between properties and predicates that needs to be emphasized. Properties are universals of some sort, and so exist independently of language. Predicates are linguistic entities that may sometimes be used to attribute properties to things. There are properties that have no names – for example, because they have yet to be discovered – and there are predicates that truly apply to things, but do not name properties. If "a" names an individual, then the sentence "Fa is true iff a satisfies F." However, it does not follow from this that there is a property of being F which a has, and which the predicate "F" might be said to name.

This is an important point, and it is one that needs to be firmly stressed. Genuine properties are universals with tropes existing in the world. They are things that exist independently of language and independently of human knowledge. Predicates, on the other hand, are parts of sentences. They are embedded in languages, and would not exist if the languages in which they are predicates did not exist. A given predicate "F" may or may not be *satisfied* by a given object a. If it is, then a sentence of the language (of which it is a predicate) ascribing that predicate to that object will be true. But still there may be no property of F-ness, or even property kind, F-ness, that the object a has. There must indeed be properties and structures of *a*, and/or perhaps properties and structures of other objects, to which *a* is asserted to be related by the claim that F*a*, in virtue of which the predicate F is satisfied by *a*. But it does not follow from this that F names a property or property kind. Mangoes are delicious. But there is no property of deliciousness that mangoes have – at least, not in my ontology.

If there is no one to one correspondence between predicates that are true of things and properties, the question arises as to which predicates designate properties? Like Armstrong (1989), I do not think that this question can be answered *a priori*. Nevertheless, there are two *a priori* principles that can be used to restrict the field of possible candidates. They are:

1. *The maxim of discovery.* If properties exist independently of language, then it cannot be the case that we can discover new properties just by the artful manipulation of language.

New properties have to be discovered in nature, not inferred from the predicates we use to describe things. The maxim of discovery obviously has important implications for the ontology of properties, some of which will be discussed later.

2. *The principle of non-proliferation.* If something has a natural property P, then the only other natural properties it logically must have in virtue of having P are just those on which P is ontologically dependent.

The principle of non-proliferation allows that the existence of one property might entail the existence of another, if this other property is one on which the given property is ontologically dependent. That something has mass *m*, for example, entails that it has mass. But this is as it should be, because the more specific property in this case (having mass *m*) is ontologically dependent on the more general one (having mass).[35]

Using the maxim of discovery, I argue that Armstrong (1989) is basically right about disjunctive and negative properties, but wrong about conjunctive properties. Conjunctive properties, even if the conjuncts are sometimes co-instantiated, may not exist. For example, the properties of being cubical and being positively charged certainly exist. But it does not follow from these facts that there is also a property of being a positively charged cube, even if there are things that happen to answer to this description. For the conjunctive universal (of being a positively charged cube) is discovered, if it is discovered at all, simply by the linguistic device of conjoining predicates naming genuine universals. Therefore, to admit

90

the conjunctive universal into one's ontology as well as the conjuncts simply on the ground that the predicates designating these conjuncts can be conjoined without contradiction is to start playing the linguistic game. It is to pretend that one can make discoveries about what there is in the world just by the artful manipulation of language.

For similar reasons, I would argue that property names that are derived from natural kind names do not name properties. Certainly, the property of being an electron (if, indeed, there were such a property) could not be an ontologically fundamental one, and the same must be true of all properties (assuming that they exist) whose names are generated from natural kind names. The property of being an electron would not be ontologically fundamental because its having instances is ontologically dependent on there being things that have unit charge, unit mass, spin 1/2, and so on. The property of being an electron would therefore have to be ontologically dependent on these other properties. Yet, electrons, protons, and the like, might well be ontologically fundamental kinds of objects. Therefore, to insist that there is a property of being an electron or one of being a proton is to turn what might well be a fundamental existent in the category of substances into a neutral thing (which, inexplicably, could not be anything other than an electron or a proton, as the case may be) with a property that is ontologically dependent on all of these other properties.

While I find this argument quite convincing, there are many who do not. A reader of the manuscript for this book, for example, has argued that electrons are nothing more than bundles of causal powers. There is no bare particular or neutral object in which these various causal powers are co-instantiated. Their tropes are simply compresent in an entity that is an electron. Given this perspective, it follows that the property of being an electron is instantiated not in, but by, an electron. An electron is a freestanding trope of the conjunctive property universal – that of being an electron.

This is certainly a plausible way to conceptualize the case. For I do not believe, and never have, in the existence of bare particulars, or featureless substances, in which the supposed properties of being an electron or being a proton might be instantiated. I have chosen to make this point by insisting that there is no such conjunctive property. However, I should be equally well satisfied with a fundamental distinction between two kinds of properties – properties that can have freestanding tropes and properties that cannot. For if the property of being an electron exists, then it is at least a very special kind of property – one that can be instantiated without being instantiated in anything. I am in complete agreement with the proposition that an electron is nothing more than its causal powers. This is not at issue. What is at issue is whether there are two kinds of universals – property universals which have no freestanding tropes and substantive universals which do.

My manuscript reader insists that substantive universals are a species of property universals. I should be happy enough to accept this proposal insofar as it applies to things such as electrons and protons, which are very plausibly just bundles of causal powers. But I am not convinced that all substantive universals can be reduced in this way to property universals, because there is as yet no satisfactory theory of the structural universals of the sort that would be needed for this task to be completed. However they may be conceptualized, there are certain substantive universals that are fundamental in the hierarchy of substances, and these universals deserve a special place in our ontology. They should not be lumped together with fabricated conjunctive universals such as "positively charged cube." Therefore, even if we accept the proposed reduction of substantive universals to conjunctive property universals, we should at least demand a sparse theory of conjunctive property universals – one that would include that of being an electron, but exclude that of being a positively charged cube.

On reflection, it seems clear enough that the property of being negatively charged is different in a number of ways from the alleged property of being an electron. Many different kinds of things, including electrons, can have the property of being negatively charged. But clearly nothing other than an electron can have the supposed property of being an electron. And even then this is not really a property that the electron has, because to have this property it seems that it would have to be something else – for example, a bare particular. I say that an electron is an electron, not something that has the property of being an electron. In my view, there is no such property. There is no property of being a horse, either, and for the same sort of reason.

The idea that there is a property of being an electron which something or other, presumably an underlying substance, might have does not belong to the conceptual framework of modern science. It belongs, rather, to the conceptual scheme of the Council of Trent, which attempted to make sense of the Catholic doctrine of transubstantiation. If you can believe in a metaphysics which allows that the "species essence" of electronhood can inhere in a substance, the identity of which is independent of the properties that constitute this essence, then I shall acknowledge your entitlement to believe that there are things that might, or might not, have the property of electronhood. But this seems to me to be just medieval mumbo-jumbo of a kind that has no place in a modern philosophy of science.

92

It is important to understand that in the view that is being advocated here, not every grammatical predication is the attribution of a natural property, even if the resulting sentence is true. Here are just some of the many kinds of cases:

1. Some predicates are evaluative, such as "...is good", "...is nasty" and "...is just," but there may be no properties of goodness, nastiness, or justice that are attributed by their use. There are no doubt natural properties in virtue of which things are, or are judged to be, good, nasty, or just. But these properties are not themselves goodness, nastiness, or justice.
2. Some predicates are used, not to attribute natural properties to things, but rather to deny that things have certain natural properties. The predicates "is not spherical" and "is not at Absolute Zero," for example, do not denote natural properties, but signify that whatever properties their subjects may have, they are not these.
3. Some monadic predicates, such as "...is under the apple tree" and "...is pre-historic" are really contracted relations, and do not denote natural properties in their own right. There are natural relations of "being under" and of "occurring at an earlier time than" that may hold of pairs of objects or events, but there is no independent natural property of "standing in the relation R to B," even though the object A may satisfy this predicate.
4. Some predicates, such as "...is legal" and "...is socially approved" apply to things in virtue of social conventions, attitudes, or practices, and tell us more about society than about the things of which they are predicated.
5. Existence is in a category of its own, and is often said not to be a property. However, "...exist" is grammatically a predicate, since it can sometimes be inserted after a proper name or a general name to make a true sentence. Whether existence is a property or not is of course arguable.
6. Some predicates that can always be truly applied to things are obvious fabrications, and there is little temptation to say that they denote natural properties. The predicate "...is such that either p or not p" is a good example. Everything is such that either p or not p, and nothing is such that both p and not p. However, nothing is to be gained by postulating the existence of properties like these.
7. Of special interest are predicates with embedded conditionals, such as "...is disposed to do Y if X occurs." Opinion is divided about whether there are natural properties (dispositional properties) denoted by such predicates. Many philosophers think there are no such natural properties. I think there are dispositional properties. The theory of dispositional properties will be taken up in Chapter 3.

In my view, not every natural property has a name and not every predicate is used for natural property attribution. Hence, there are natural properties without corresponding predicates and there are predicates that are satisfied by objects but do not attribute natural properties to those objects. The same is true of n–adic relations and predicates. Not every natural n–adic relation can be named and not every n–adic predicate that is satisfied by a group of n things is to be construed as the attribution of a natural n–adic relation to these things. Predicates that are attributive of natural properties or relations will be said to *denote* or *designate* the natural properties or relations they attribute, although the relation between a predicate that denotes a natural property and the property it denotes is not straightforwardly one of naming or denoting.

The distinction between predicates and properties is important for several reasons. First, if the only properties or relations that existed were those we were able to name, if not in English, then in some other language, then the ontology of properties and relations would be radically dependent both on language and on human experience. But what there is is not thus dependent. Moreover, if truly applied n–adic predicates always implied the existence of corresponding n–adic natural properties or relations, then the distinction between natural properties and relations, on the one hand, and predicates, on the other, would be of little use. For it would not help us in any way at all to understand what makes a true sentence involving a given predicate true. For example, if *a* satisfies "F," then one reason for this might be that it does so because it has the property of being F. Another might be that it does so because anything that is either G or H satisfies "F." But if *a*'s satisfying "F" entails its having the property of being F, then the explanation disappears.

Second, predicates can be elaborated and operated on in many different ways to produce new predicates. And, in many cases, these predicates will be truly applicable to a given object if and only if the predicate with which we began is truly applicable to that object. But if truly applicable predicates can be so easily invented, it is hard to believe that they all denote real properties. What exists in reality has to be discovered, not created just by the artful manipulation of language.

To deal with this problem, we appeal to the *principle of non-proliferation:* Suppose, contrary to the principle of non-proliferation, that satisfaction of a predicate is a sufficient condition for the existence of a real property distinguished by that predicate. Then, since any red object *a* satisfies the predicate "... is red," the existence of this object implies the reality of the property of redness. So far, so good, if there really is a property of redness. However, by the same argument, *a* must have the property of being red or yellow, and of being red or shiny, and generally of being red or ..., whatever other disjunct we may care to insert. For clearly *a* must satisfy all such disjunctive predicates. It is also clear that *a* must satisfy the negative predicates "... is not blue," "... is not yellow," and so on. Therefore, *a* must have all of these negative properties and of course all of the disjunctive properties that can be generated from these negative properties too. Moreover, if *a* has the property of being red – if indeed this is a property – then *a* must have the second-order property of having the property of being red and the third-order property of having the property of having the property of being red, and so on. And of course there will be a similar infinite set of higher-order properties for each of the predicates generated from the base predicate "... is red" by disjunction or by negation of contraries. I opt for non-proliferation. I say that if redness were indeed a genuine property (which it is not), then there would be only one real property here[16] – namely, the property of redness.

The principle of non-proliferation implies that there are no disjunctive properties whose disjuncts are real properties. For it follows from this principle that there could not be a real property of being FvG, unless the properties of being F or of being G were ontologically dependent on it. But this is impossible. Therefore I follow Armstrong (1978) in claiming that there are no real disjunctive properties.[17]

What about so-called *negative properties?* If "F" is any predicate denoting a property, can the negation of "F" also denote a property? It depends on what kind of property it is. *Active properties* – for example, properties of the nature of causal powers, capacities, or propensities – do not have negatives. Where two predicates "F" and "G" are contradictory, not more than one of them can denote a real property of this kind. *Passive properties,* on the other hand – for example, those usually described as categorical – may have negatives. The properties of being open and being closed, for example, would both appear to be genuine properties.

This answer accords well with our intuitive judgement that if "F" denotes a causal power, then "–F" does not also denote a causal power. It merely signifies the absence of the power denoted by "F." However, there is a difficulty with this suggestion – namely, which, if either, of a given pair of contradictory predicates denotes a causal power and which merely signifies its absence? The question is one of ontology, so it cannot be settled linguistically. For example, it will not do to say that the predicate that contains no negation operator is, if either of them is, the one that denotes a causal power. For even if this were always the case, we should still want to know why it was. What sometimes makes one of two contradictory predicates a real property designator and the other not?[18]

95

The answer lies in the causal roles of these kinds of properties. For the judgment concerning which (if any) of two contradictory predicates denotes a real property is one about the causal roles of the alleged properties. The matter is complicated, however, because our judgments about causal roles are sometimes confused. If you hit a golf ball that lands on the green, there is no doubt about the direction of causation. But suppose you pull the curtain and darken the room. Have you then caused the room to darken? Yes, you might say, but you have done so only by removing the cause of its being light. By pulling the curtain, you have reflected the light back to where it came from, and the room has become dark by default.

Let us therefore distinguish negatively acting causes such as pulling the curtain from positively acting causes such as hitting the golf ball. Theoretically, I suppose, the distinction will have to be drawn by reference to the direction of energy transfer in the causal process. For positively acting causes, the direction of energy transfer is from cause to effect. For negatively acting causes, the effect is produced by depriving the object affected of energy it would otherwise have received. Negatively acting causes produce their effects by default.[19] Of course, this distinction can only be applied usefully in the simplest cases. In more complex cases, there will often be a mixture of positive and negative actions.

The active properties of a thing are the sources of positive causal action by that thing. If redness were a real, active property, for example, then it would be so because red things have a certain active capacity—for example, the capacity to reflect certain kinds of light.[20] Accordingly, not being red cannot also be a real active property. This is not to say that the "negative property" of not being red cannot feature in a causal explanation. We might, for example, fail to stop at a stop sign because someone had forgotten to paint it red. But its not being red is a negatively acting cause in this case, and so its causal role is different from that of a genuine active property.

What about *conjunctive properties?* There is no case against conjunctive properties to be derived from the principle of non-proliferation. Nor is there any case to be derived from my rejection of negatively acting properties. On the contrary, conjunctive properties, it would seem, are always ontologically dependent on their conjuncts. Nevertheless, the grammatical device of conjoining predicates is just that – a grammatical device. We could get along perfectly well without this device. To say that the billiard ball is both round and shiny is just to say both that the billiard ball is round and that the billiard ball is shiny. Thus, predicative conjunction is just shorthand for propositional conjunction. Therefore it is merely an accident of language that we are grammatically able to conjoin predicates in this way.

Hence, by the maxim of discovery, the fact that the grammar of English allows us to conjoin predicates gives us no good reason to believe in conjunctive properties. To show that there is a property of being both round and shiny, we should need an argument that this conjunction of properties has a distinctive causal role.

As we have seen, there is a case for regarding the property of being an electron as a conjunctive property, the conjuncts of which are the properties of electrons. However, if there is such a property as this, then it is a very special kind of property. For it is one that has freestanding tropes, and is thus instantiated by objects (namely, electrons), not in them. I call them "substantive universals."

I therefore part company with Armstrong (1990) on conjunctive properties. If any of them is real, then it is because there are intrinsically significant conjunctions of properties that we are forced to recognize as existing in nature, independently of language. That is, they must be conjunctions of properties that require some explanation better than their accidental conjoinability.[21] For reasons to be elaborated in Chapter 3, I also part company with Armstrong on dispositional properties – properties that are normally characterized by subjunctive conditionals.

2.9 SO WHAT ARE NATURAL KINDS?

Natural kinds, as I understand them, are generalized universals. They are generalized in two ways, (1) in respect of *category*, and (2) in respect of *position in a hierarchy*. Classical universals are universals in the category of properties and relations. As such, they are instantiated in objects or substances, not by them. Generic universals of this classical kind (for example, quantities) are natural kinds whose infimic species (for example, specific quantities) are classical universals. Generic universals of this sort are what I call "property kinds." The instances of property kinds are the tropes of these kinds of properties or relations occurring in the objects in which they are instantiated.

Substantive universals belong to a different category from classical universals. For they are instantiated, not in objects, but by objects, and the instances of substantive universals are not tropes, but the objects by which they are instantiated. The species electrons, protons, and neutrons are all substantive universals. Substantive natural kinds are then just generic substantive universals. The baryons, for example are a natural kind, of which protons and neutrons are infimic species.

Natural kinds of processes are different again. They are the natural kinds occurring in the category of events and processes, which have events and processes as their instances. Natural kinds are not classes of things, although there are things that we say are instances of natural kinds. They are not complex properties or structures, although the instances of a natural kind may be characterized by certain properties or structures. The instances of any given natural kind are necessarily intrinsically the same in some respects, and the respects in which they are necessarily the same are the essential properties or structures of the kind. These are the properties or structures in virtue of which these things are things of the kinds they are.

All classical universals are species of natural kinds. They are the *species infima,* or ultimate species, of property universals. The species infima of substantive and dynamic kinds are formally very much like classical universals, except that they have different kinds of instances. Classical universals, because they are at the end of the line, cannot have species. The instances of classical universals must therefore be identical. It is not necessary to say in what respect they must be identical, for if their instances have respects, they must be identical in every respect. This is why it is so important to distinguish between the property trope that instantiates a classical universal and the object in which it is instantiated. For it has to be possible for the same classical universal to be instantiated in things that are otherwise very different. For the infimic species of substantive and dynamic universals, the requirement is that their instances be essentially the same. If X is such a substantive or dynamic universal, then every instance of X must be essentially the same as every other instance of X.

As generalized universals, natural kinds permit ranges of intrinsic variation. In the case of spectral kinds, this range is a continuous one. The instances of natural kinds may thus differ from each other incidentally. They are united by the fact that they all have the essential properties of the kind. They are distinguished by the fact that they have the distinctive properties of more specific kinds. In general, the instances of a natural kind that are intrinsically and non-accidentally different from each other are necessarily instances of different species of the kind. Thus, natural kinds speciate, and the process of speciation necessarily ends with universals.

The only universals that are generally recognized in the literature are property (or relation) universals. It is argued here that there are at least two other sorts of universals – *substantive* and *dynamic.* Substantive universals are the ultimate species of substantive natural kinds. Dynamic universals are the ultimate species of dynamic natural kinds. The different sorts of natural kinds and universals are distinguished from each other by the natures of their instances. The best arguments in favor of this conception are

(a) that it makes a neat and comprehensive system, and (b) there is no other metaphysics that comes anywhere near to dealing as adequately with the known structure of reality. Philosophers have spent far too long, and succeeded far too little, in trying to use the classical Aristotelian concepts to do a twenty-first century job.

As we have seen, natural kinds are not ontologically dependent on their species, and hence not on the universals that are their ultimate species. On the contrary, they are at least as fundamental in ontology as any of their species. Hence, physical quantities and other natural property kinds and generic kinds of natural processes are not high-level constructs, but ontologically basic entities.[22] Therefore they are bound to be of primary concern in the natural sciences.

Consider finally whether natural properties define substantive natural kinds. I think not. Natural properties are objective, mutually distinct, and satisfy the intrinsicality and essentiality requirements on natural kinds. However, natural property classes do not satisfy the speciation or hierarchy requirements.[23] They do not speciate as natural kinds classes must. An electromagnetic wave and a sound wave, for example, may both have wavelength l. But an electromagnetic wave and a sound wave are not the same kind of thing, even if they do have the same wavelength. They are not species of a natural kind that consists of things having wavelength l. There is no such natural kind. Nor are the property classes to which any given object belongs necessarily related hierarchically. The class of negatively charged objects is not a species of objects with mass greater than four grams, or conversely – although obviously – an object can have both of these two natural properties.

If natural property classes – classes defined by natural properties – are not necessarily the classes of substantive natural kinds, then what are substantive natural kinds? Clearly, if the natural property that defines a given property class happens to be the conjunction of all of the essential properties of a certain natural kind K, then the class and the natural kind must have the same extension. For a thing can be a member of this class if and only if it is a thing of the kind K. However, there is no further property, over and above this conjunction of essential properties, that is the property of being a thing of the kind K. For a thing's being of the kind K is ontologically dependent on, indeed it is ontologically reducible to, its having this conjunction of properties, not the other way round. Therefore, by the principle of non-proliferation, we need a separate argument for the existence of a property of being a member of a given natural kind. However, there seems to be no such argument.

From what has been said informally about natural kinds, it seems that the following theses are necessarily true:

1. If $x =_i y$, then $x =_e y$.
2. For every K, there is an intrinsic property P such that PeK.
3. If $x \in$ K and PeK, then $\Box$ Px.
4. If PeK and $K_1 <$ K, then PeK$_1$.
5. If $x \in K_1$ and $K_1 < K_2$, then $x \in K_2$.
6. If $x, y \in$ K, and $x \neq_e y$, then there is a K_1 and K_2 where $K_1 \neq K_2$, such that $x \in K_1$, $y \in K_2$, and $K_1, K_2 <$ K.
7. If $K_1 \neq K_2$, then there is a property P such that $\sim$ (PeK$_1 \equiv$ PeK$_2$).
8. If $K_1 < K_2$, and $K_2 < K_3$, then $K_1 < K_3$.
9. If $x \in K_1, K_2$, and $K_1 \neq K_2$, then either $K_1 < K_2$, or $K_2 < K_1$, or there is a K such that $K_1, K_2 <$ K.

Where, '$x \in$ K' = 'x is an instance of the natural kind K'

'PeK' = 'P is an essential property of K'

'$K_1 < K_2$' = 'K_1 is a species of K_2'

'$x =_e y$' = 'x is essentially the same as y'

'$x =_i y$' = 'x and y are intrinsically identical in their causal powers, capacities, and propensities'

To these we may now add the following:

10. For all x, ($\Box x \in$ K or $\Box x \notin$ K).
11. There are no two natural kinds, K_1 and K_2, such that *necessarily*, for all x, $x \in K_1$ or $x \in K_2$.
12. The class of things defined as the intersection of the extensions of two distinct natural kinds K_1 and K_2 is not necessarily the extension of a natural kind, unless $K_1 < K_2$, or $K_2 < K_1$.
13. The class of things defined as the union of two distinct natural kinds K_1 and K_2 is not necessarily the extension of a natural kind, unless $K_1 < K_2$ or $K_2 < K_1$.

What 10 says is that whether something is an instance of a natural kind is never a contingent matter. If x is an instance of a natural kind K, then it is so necessarily. If it is not an instance, then this is not just accidental. It could not be or become an instance. Thesis 11 affirms that if K_1 is a natural kind, then there is no complementary natural kind K_2 (= not–K_1), although the class of things that are not K_1 might *accidentally* be the extension of a different, and perhaps wholly unrelated, natural kind K_2. That is, there are no negative natural kinds. If the class of ravens is a natural kind, then the class of non-ravens is not. In this respect, natural kinds are like natural objects. If Brian Ellis is a natural object, then non-Brian Ellis is not also a natural object (although of course it could be the case that there is only one other natural object in the universe). Theses 12 and 13 assert that the intersections and unions of natural kind classes need not also be natural kind classes, unless one of these classes is a species of the other.[24]

We suppose these principles to be logically necessary. However, the kind of necessity involved within these principles is generally metaphysical. What is necessary metaphysically is, like what is logically necessary, true in all possible worlds. But it is not true under all interpretations of the non-logical terms. Moreover, what is metaphysically necessary is not *a priori*. That is, it has to be discovered by empirical investigation.

NOTES

1. One of the best discussions in the literature of the question whether there are hierarchies of natural kinds is to be found in Elder (1994a, 258). Elder argues that there are indeed such hierarchies. He speaks of "middle-level" kinds and of "subordinate" and "super-ordinate" kinds. The problem, as he sees it, is to distinguish genuine super-ordinate kinds from disjunctions of kinds and sub-ordinate kinds from mere variants, or versions, of middle-level kinds. He argues conclusively that the "individuation of kinds, over which competition between any two properties F and G is defined, must be multiply grounded. The kinds from which F would exclude G and G would exclude F must be distinct from one another independently of this particular exclusion; and since kinds can be distinct from one another only by being differently characterised, the kinds must independently make selections from other ranges of competing properties."

2. Objects should not be thought of as located within fixed material boundaries in space and time, as atoms were once thought to be, but as having a kind of sphere of causal influence that extends well beyond any such boundaries, rather in the manner of a Boscovichian atom. For more on this, see Section 8.3.

3. Perhaps this is what Aristotle had in mind in his thesis *Z6*. "... each primary and self-subsistent thing is one and the same as its essence" (*Metaphysics Z*, Ch. 6, 1032a 5–6). However, I do not pretend to be an Aristotle scholar, and I happily leave the interpretation of *Z6* to the experts. The thesis is discussed in Code (1986), where it is used to construct a logic of being and having, which is precisely the distinction I am trying to make.

4. Donnellan (1983) argues that the fact that gold and other elements are identified by their atomic numbers rather than by their atomic weights is plausibly a fact about our science reflecting the relative abundances of isotopes in our world. On Twin-Earth, where the relative abundances are different, they might well deny the fact that gold has atomic number 79 expresses a necessary proposition. Indeed they

might. But if they were to do so, they would be wrong, as the scientists on Twin-Earth would eventually discover. For the necessity of "gold has atomic number 79" is a real or metaphysical necessity. It does not depend on the conventions of language either here, or on Twin Earth, or anywhere else.

5. Fales (1990) uses the term "generic universals" to refer to what I am here calling "generic kinds." His explanation of the genus/species relation, although expressed in terms of causal relations, appears to be the same as mine. He says: "If S and S' are specific universals and G is a determinable or generic universal under which they both fall, then the causal relations between G and other universals are a subset of the causal relations between S and other universals, and also a subset of the causal relations between S' and other universals" (p. 239).

6. John Bigelow disagrees with me about this. The nature and reasons for his disagreement are discussed in Section 7.5.

7. We speak of the *relevant subvenient basis* or *RS-basis* of a property rather than its *causal basis* (see Prior, Pargetter, and Jackson, 1982) because it is important to distinguish between the RS-basis of a causal disposition and the causal antecedent of such a disposition.

8. Given this definition, it is clear that P is not ontologically dependent on the supposed disjunctive property PvQ even though P could not be instantiated unless PvQ were also instantiated. For instantiation of PvQ does not entail instantiation of P. It should also be clear why being red is not ontologically dependent on being not blue. For being not blue is not a sufficient condition for being red. The principle of non-proliferation therefore requires us to deny that these disjunctive or negative properties exist.

9. Fine (1995) considers two basic accounts of ontological dependence – one in terms of modality and existence (ME), the other in terms of essentiality and existence (EE). His definitions are: ME: One thing x depends ontologically on another y just in case it is necessary that x exists only if y exists. EE: One thing x depends ontologically on another y just in case it is an essential property of x that it exists only if y exists (p. 272) Fine explains EE: The essence of x is the collection of propositions that are true in virtue of its identity (p. 275). Fine then refines EE to make it relative to constitutive essences, where: "A property belongs to the *constitutive* essence of an object if it is not had in virtue of being a logical consequence of some more basic essential properties; and a property might be said to belong to the *consequential* essence of an object if it is a logical consequence of properties that belong to the constitutive essence" (p. 276). Toward the end of the paper, Fine considers type/token or generic dependencies. Here his theory strikes trouble, for "it is presumably true that a token cannot exist without a type of that token also existing. But we do not want to say, on that account, that a token also depends on one of its types" (p. 288). But if the type is a natural kind, why shouldn't the dependence be upward from the token to the type? I think it is.

10. The distinction between red and orange, for example, is a distinction that depends on how we see the world – that is, it depends on our color vision, and does not exist in the world independently of the ways in which we process color information. But of course I am not denying that there are real, intrinsic differences between differently colored things. It is just that the way we draw the boundary

between red and orange is not mandated by these real, intrinsic differences. In reality, there is just a continuum.

11. Quinton's (1957) natural class theory is the view that all genuine properties and relations are ultimately reducible to natural classes. To illustrate: There is a natural class of electrons on which the property of being an electron supervenes and there is a natural class of negatively charged things on which the property of being negatively charged supervenes. The natural class of electrons is included in the natural class of negatively charged things. Hence, electrons are negatively charged, or, as we commonly say, they have the property of being negatively charged. I think that Quinton is right about the property of being an electron, but wrong about the property of being negatively charged, and wrong about why electrons have this property. Armstrong (1989) describes Quinton's theory as "class nominalism." He objects to it on the ground that one cannot distinguish satisfactorily between contingently co-extensive properties on Quinton's theory – unless one has natural classes that extend across all possible worlds. To Armstrong's objections, we should add the following: If we allow natural classes to range across possible worlds, then we need to know a lot more about these classes and why they have the properties they do. For example: Do natural classes exist necessarily? If not, is there a possible world in which there are no natural classes – that is a world without properties or relations? Could two worlds consisting of intrinsically identical objects be divided into different natural classes, and so be made up of things that bear a whole lot of different extrinsic properties and relations? If the identity of a natural class defines what we think of as its defining property, then what gives unity to the members of that class? What makes it a natural class, as opposed to any other kind of class? Nevertheless, a variant of Quinton's theory might still be defensible. For it is plausible to suppose that the properties that are of most interest in science are something like natural classes of causal powers. There are, for example, two natural and distinct classes of causal powers relating to the properties of unit charge and unit mass. And these would be different, even if these two properties happened to be co-extensive. Indeed, they would be different even if they were necessarily co-extensive, because they would feature in the causal explanations of different phenomena. Hence these two properties might be seen as being dependent, not on the natural classes of things which possess them, but on the natural classes of causal powers they summarize.

12. Shoemaker's (1980) criterion of property identity is: ". . . if under all *possible* circumstances properties X and Y make the same contribution to the causal powers of the things that have them, X and Y are the same property" (p. 114). This is a much more satisfactory theory of property identity than necessary co-extensionality, because it derives from his theory of what properties are.

13. Here I am in agreement with Shoemaker (1980). See note 12.

14. This is argued in Ellis (1976).

15. See Section 2.6.

16. Redness is not a real property, since our color classifications are not natural classifications.

17. This is one reason why I find Hardegree's (1982) logic of natural kinds unsatisfactory. For neither are there any real disjunctive natural kinds.

18. Phil Dowe (1999) has some useful things to say on this question.

19. Of course, energy transfers often occur in both directions, with neither one predominating. In all such cases, we should speak of causal interactions, and the effects as the results of these interactions.

20. Although, for reasons already given, we do not think that redness is a natural property (or property kind).

21. We also part company with Armstrong (1968) and other so-called "categorical realists" – for example, Prior, Pargetter, and Jackson (1982) and Prior (1985) on dispositional properties – that is, properties that are normally characterized by subjunctive conditionals. See Ellis and Lierse (1994a).

22. This is one reason why it is wrong to try to construe natural kinds as structural universals. For if they were, then they would be ontologically dependent on the universals from which they were constructed. But this, as we have seen, gets the relation of ontological independence the wrong way around.

23. Here I find myself in agreement with C.S. Peirce and J.S. Mill. I. Hacking (1991) reports Peirce as advocating the following definition: "Any class which in addition to its defining character has another that is of permanent interest, and is common and peculiar to its members, is designed to be preserved in that ultimate conception of the universe at which we aim, and is accordingly to be called 'real'" (p. 119, quoted from "Kind," in Baldwyn's *Dictionary of Philosophy and Psychology*, 1903, **1**, p. 60f). Commenting on J.S. Mill's distinction between natural property and natural kind classes (p. 121), Hacking says that members of the one type of class share a single property, while members of the other type of class share a manifold of properties. "White things," Mill wrote, "are not distinguished by any common properties, except whiteness: or if they are, it is only by such as are in some way connected with whiteness. But a hundred generations have not exhausted the common properties of animals or plants, or sulphur or phosphorus; nor do we suppose them to be exhaustible, but proceed to new observations and experiments, in the full confidence of discovering new properties which were by no means implied in those we previously knew."

24. Many of these principles are distinctive natural kinds vis à vis other sorts of classes. Even the principles concerned with the species relation are distinctive. Hardegree (1982) argues that natural kinds form a lattice structure. Specifically, he defines a natural kind to be an ordered pair consisting of a set A of individuals and a set T or traits, subject to the requirement that T determines the set A of individuals having all of these traits, and the set A determines the set T of traits possessed by all members of A. These two conditions, he says, imply that both A and T are Galois closed, and that every member of A instantiates every trait in T, and that no individual not in A instantiates every member of T (see p. 126). This definition is satisfactory as far as it goes. But it fails to capture a number of important properties of natural kinds and their memberships. For example, there are no axioms or theorems in Hardegree's system corresponding to axioms 9, 11, 12, or 13. Hardegree defines relations that he calls "implication," "conjunction," and "disjunction," following common lattice theory practice. The implication relation, symbolized by "is the lattice theoretic version of the "species" relation. That is, "$K_1 \leq K_2$" may be read as "K_1 is a species of K_2." The disjunction of K_1 and K_2, $(K_1 \lor K_2)$, is the smallest genus of which both K_1 and K_2 are species, and the conjunction of K_1

and K_2, $(K_1 \& K_2)$ is the largest species that is a species of both K_1 and K_2. Given these definitions, we have the following theses (see p. 130):

1. $K \leq K$.
2. if $K_1 \leq K_2$, and $K_2 \leq K_3$, then $K_1 \leq K_3$.
3. $K_1 \leq K_2$ and $K_2 \leq K_1$, only if $K_1 = K_2$.
4. $K_1 \& K_2 \leq K_1, K_2$.
5. $K \leq K_1$, and $K \leq K_2$, then $K \leq K_1 \& K_2$.
6. $K_1, K_2 \leq K_1 \vee K_2$.
7. if $K_1 \leq K$, and $K_2 \leq K$, then $K_1 \vee K_2 \leq K$.

Hardegree argues that there are no other principles satisfying natural kind lattices. Thomason (1969) had suggested that natural kind lattices are also modular – that is, that two kinds overlap only if one is a species of the other. However, examples in chemistry make this thesis implausible. Nevertheless, it is plausible that there is a unique supremum in each ontological category, but no unique infimum – that is, the lattice for each category is closed above, but open below.

3

Powers and Dispositions

Scientific essentialism presents a view of reality that stands in sharp contrast to the more widely accepted Humean ones. The scientific essentialist holds (a) that matter is not passive, but essentially active and reactive, (b) that the essential properties of things belonging to natural kinds include dispositional properties – that is, causal powers, capacities and propensities, (c) that the basic laws of nature are not descriptions of behavioral regularities, but of the ways in which things belonging to natural kinds must be disposed to act or interact, given their essential properties, (d) that the causal laws of nature are metaphysically necessary, because anything that has the essential dispositional properties of a given natural kind must be disposed to behave as these properties require, and (e) that elementary causal relations involve necessary connections between events – namely, between the triggers and displays of basic dispositional properties.

The ontology required to defend these theses includes various causal powers, capacities, and propensities as ontologically basic properties. That is, it is an ontology in which things may be supposed to have such properties intrinsically. But these properties are all dispositional in character, and traditional wisdom has it that dispositional properties cannot be ontologically basic. If anything has such a property, it is said, then it must have it in virtue of its categorical (non-dispositional) properties, and as a consequence of what the laws of nature happen to be. But, it is argued, any properties that thus depend on what the laws of nature are cannot be intrinsic. Moreover, if they cannot be intrinsic properties, then they cannot be essential properties either. For, in another world, where the laws of nature are different, things that are intrinsically just the same might lack some

106

or all of these properties. Therefore, it is argued, the causal powers, capacities, and propensities of things cannot be among their essential properties, ie. their kind-identities cannot depend on them.

3.2 THE DEAD WORLD OF MECHANISM

From the perspective of seventeenth- and eighteenth-century mechanism, the objective world is not intrinsically active. It is a world, according to Burtt (1932), that is "hard, cold, colourless, silent, and dead; a world of quantity, a world of mathematically computable motions in mechanical regularity" (p. 237). Descartes, Locke, and Newton certainly believed something like this, as did most of their eighteenth-century followers. For Descartes, the essence of matter was just extension. It occupied space, and therefore had essentially only the attributes of things vis-à-vis their extension in space – shape, size, and so on. For Boyle, Locke, and Newton, the qualities inherent in bodies were just the primary qualities – number, figure, size, texture, motion and configuration of parts, impenetrability, and perhaps also body (or mass). If things with the same primary qualities were nevertheless different, then this difference must be due to differences in the primary qualities of, spatial relations between, or motions of, their elementary parts.

The qualities by which things are known to us are the qualities of experience – that is, their color, taste, warmth, odor, feel, and so on. These qualities are known to us by the sensory ideas to which they give rise. Locke calls the powers that produce these sensory ideas "the secondary qualities." According to Locke (1690), these powers are not really inherent in the objects as they are in themselves. In themselves, the objects of experience have only the primary qualities. Nor can the sensory ideas be supposed to resemble in any way the powers of the objects to induce them in us. For these powers must be supposed to be grounded solely in the primary qualities of the insensible parts of these objects, which are of an altogether different character from any of the ideas they furnish.

Locke distinguished two kinds of powers – active and passive. The active ones are the powers of things to make changes; the passive ones are the abilities of things to receive changes (p. 234). God, he supposed, had only active powers. Inanimate things, he speculated, may have only passive ones. If this is right, then created spirits, such as ourselves, would be the only things to have both active and passive powers. When we exercise our free will in some voluntary action, we certainly display an active power, according to Locke. Hence, there is no doubt, he thought, that human beings, qua created spirits, have active powers. But also, when we perceive anything, we display our capacity to be affected by it. So it is evident that qua created spirits, we also have passive powers.

The question of importance in the present context is whether active powers exist in inanimate nature. What Locke believed about this is a question of scholarship that need not concern us. But certainly a great many seventeenth- and eighteenth-century mechanists did believe in the complete passivity of inanimate nature. If one object seems to affect another – for example, crashes into it and so causes it to move, then what is involved is not so much an action on the part of the first body as a passion. As Locke (1690) explained:

A Body at rest affords us no *Idea* of any *active Power* to move; and when it is set in motion it self, that Motion is rather a Passion, than an Action in it. For when the Ball obeys the stroke of a Billiard-stick, it is not any action of the Ball, but a bare passion: Also when by impulse it sets another Ball in motion, that lay in its way, it only communicates the motion it had received from another, and loses in it self so much, as the other received; which gives us but a very obscure *Idea* of *active Power,* which reaches not the Production of the Action, but the Continuation of the Passion (p. 235).

Perhaps the mathematician Leonhard Euler adequately represents mid-eighteenth century views on causal powers. In his *Letters to a German Princess* (1795), composed in the early 1760s, he addressed at length the question of what kinds of powers exist in the world, and what their sources are (vol. 1, 295–340). He argued, as Locke had speculated, that the powers existing in inanimate nature are all essentially passive. Indeed, he thought that the powers necessary for the maintenance of the changing universe would turn out to be just the passive ones of inertia and impenetrability. There are no active powers, he argued, other than those of God and living beings. Consequently, if the mechanist's world-view is correct, the myriad changes that we see occurring around us must all be consequential upon the inertial motions of things, and their mutual impenetrabilities. The so-called forces of nature – for example, of gravitational attraction – may describe the ways in which things are disposed to behave vis-à-vis each other. There is no doubt that things are disposed to accelerate toward each other as the laws of gravity and motion require. But the source of that disposition, Euler argued, is not an attractive force emanating from the bodies, or just a natural tendency of bodies to move according to the dictates of some pre-established harmony, as Leibniz believed, but an impulsion of one thing toward another produced by some kind of tension in the ether. When the nature of this process is fully understood, Euler supposed, the planetary motions, and gravitational accelerations generally, would all be seen to be the passive consequences of inertia and mutual impenetrability.

Plausible as some of the mechanists' arguments for this conclusion may have been, it is to be argued here that this is a radically incorrect view of the nature of reality. The real world is essentially active and interactive. It is not passive, as the old mechanists and the neo-mechanists of today believe. It is dynamic. And its dynamism stems from the existence of genuine causal powers in things, both active and passive. Locke, Euler, and the other mechanists of the period all believed in the essential passivity of nature. But they were wrong, I shall argue. The inanimate world is not passive, as they believed. Material things do have causal powers, which, in appropriate circumstances, they will exercise; and these causal powers are real occurrent properties of the things in question.

Scientists today certainly talk about inanimate things as though they believed they had such powers. Negatively charged particles have the power to attract positively charged ones. Electrostatic fields have the power to modify spectral lines. Sulfuric acid has the power to dissolve copper. The question we have to consider in this chapter is: What is the source of these powers? The old mechanist view was that things do not themselves have causal powers. The powers lie outside them. They are contained in the forces that act externally on things to change their states of motion or aggregation.

Caroline Lierse and I think that this is fundamentally wrong. Things do have causal powers, we say, and these powers are among the properties that things, at the deepest levels of existence, have essentially, without which they would not be the things they are.

3.3 FORCES AS EXTERNAL TO OBJECTS

One thing that the mechanists of the seventeenth and eighteenth centuries all had in common was their belief in the externality of forces. All changes, it was believed, were ultimately changes brought about by motion. The elementary parts of things were thought to be rigid and unchanging. So all changes must consist of changes in the arrangements or motions of the most elementary things. Forces were postulated as the links between these things to explain the changes that take place. The forces do not, of course, change the elementary things on which they operate, only how they move or are arranged. Consequently, the identities of the elementary things were considered to be independent of the forces that operate on them. Change the forces, or change the laws of nature so that new forces may come to act between things, and the same elementary things will be disposed to behave in different ways. The dispositions of things must therefore all depend ultimately on the underlying structures of the elementary things of which they are composed, and on the laws of nature that determine what forces there are, and how they operate on these most elementary things. This, with perhaps a few concessions to modernity,[1] is the doctrine known as *categorical realism*.

The identities of things were also supposed to be independent of any forces they may generate. Indeed, the most widely accepted European view seemed to be that inanimate things could not generate any forces at all. For that would imply that they had active powers, which, by their inanimate nature, they could not possess. Yet things do at least appear to have some active powers, and various kinds of forces (gravitational, electric, and magnetic) were recognized. Consequently, the natural philosophers of the period all used the language of active causal powers quite freely in their descriptions of inanimate nature, even though they believed that these powers were ultimately not active but passive. If pressed, they would say that the powers were not really inherent in the objects that seemed to possess them, but were dependent on their ultimate constitutions, and on the laws of nature that were universally supposed to be external to them.

For these reasons, causal powers, and forces generally, were regarded as occult. Hume went so far as to deny that there existed in nature anything other than the regular patterns of behavior that explanations in terms of forces were intended to explain, and when we speak of causes, he said, it is really only to such regularities that we can be referring. And it is for these reasons, if for no others, that propositions attributing causal powers to things have long been regarded with suspicion. This suspicion applies not only to active causal powers (those that are not obviously dependent on the actions of God or man), but also to the passive ones. For the two go together. For every passive causal power – the power to receive change, which is ever exercised by anything – there must be an active causal power – the power to make change, to which it is responding. Consequently, if one kind of power is suspect, then so is the other. If the power to produce a change is no more than an invariable disposition of something to behave in a certain way (or range of ways) in certain circumstances (or range of circumstances), then the power to receive change can be no more than an invariable disposition of something to respond in a certain way (or range of ways) in these circumstances (or in some range of cir-

110

cumstances). But such invariable dispositions are not thought to be real properties of the things in question. The real properties are just the underlying structures to which the laws of nature may be supposed to apply.

3.4 THE DUBIOUS STATUS OF DISPOSITIONS[2]

Dispositions, Mellor once remarked, are as shameful in many eyes as pregnant spinsters used to be, "ideally to be explained away, or entitled to a shotgun wedding to take the name of some decently real categorical property" (Mellor, 1974). This "Victorian prejudice" against dispositions still exists. On reflection, this fact is perhaps not really surprising. Dispositions have dallied in most corners of the metaphysical arena, mixing unashamedly in the (not always respectable) company of behaviorism, counterfactuals, induction, nomic necessity, and causation. For dispositions to be taken seriously, it is argued, they need to ground themselves in some decent categorical bases, for only then can their claim to be genuine properties be respected.

In this chapter, we shall argue that the subordinate status thus assigned to dispositions is unwarranted. It is a legacy, we believe, of the traditional metaphysics of the seventeenth and eighteenth centuries. If there are no causal powers, capacities, or propensities in nature, as most philosophers then held, and many philosophers still hold, then the dispositions of things cannot be explained with reference to such properties. To explain a disposition, on this view, it is necessary to describe the kinds of underlying structures that give rise to the disposition (its supposed categorical basis) and to cite the laws of nature that apply to such structures.

We do not seek to restore the reputation of dispositions by attacking the status of categorical properties, or by arguing that all properties are really dispositional, as some philosophers have done.[3] For the present lowly status of dispositions is not merely a function of their relationship to categorical properties. Rather, it is a consequence of their traditional affiliation with an inadequate ontology, based on a Humean metaphysic, and a flawed semantics of dispositional terms. What is needed, and what we seek to provide here, is a more adequate semantics, and an ontologically more satisfactory theory of, dispositions − one that allows at least some dispositions to be counted as the direct expressions of genuine occurrent properties. Such properties, we suppose, include all of the causal powers, capacities, and propensities of things that exist fundamentally in nature and which have no bases of the kinds sought by categorical realists.

In the analysis to follow, we shall follow the practice of using the term "disposition" as a general term for referring to any tendencies, abilities, liabilities, and so on that are normally called "dispositions," whether or not the predicates used to ascribe these dispositions refer to genuine properties. For, like Armstrong (in Armstrong, 1989), and in many other places, we wish to distinguish clearly between properties and predicates. To say that x has a disposition D (for example, to E in circumstances C) is to say something which, if true, is to say only that "D" is satisfied by x. It is not to say that there is a property named by "D" that is this disposition. Of course, there must be properties of x in virtue of which x satisfies "D," but there may well be no such thing as a property of D-ness. Indeed, if the categorical realists are right, there never is such a property.

We do not agree with the categorical realists about this, although we accept their distinction, and also a sparse ontology of properties roughly of the sort advocated by Armstrong. But we think that dispositional predicates sometimes do name dispositional properties. The causal powers, capacities, and propensities of substances described in scientific textbooks are mostly genuine properties whose essential natures are dispositional. We call all such genuine properties "dispositional properties." So, for us to talk of dispositional properties is to talk of real, occurrent properties, not fictitious constructs of language.

A number of philosophers have argued in favor of a non-Humean ontology that includes such real and irreducibly dispositional properties.[4] Others have suggested a realist semantics of the kind we shall advocate for terms denoting such properties.[5] What is new in our theory is the link we establish between real dispositional properties and natural kinds of processes. There are natural kinds of *processes*, we say, and among the processes which belong to such kinds, there are some which are essentially the displays of natural dispositional properties. For a process to be a member of a natural kind such as this, it must be a genuine display of the appropriate dispositional property.[6] If it is not, then it is not truly an instance of this kind of process, however like it it may appear to be. The laws of nature which describe these kinds of processes are thus related directly and essentially to *natural dispositional properties*. The natural dispositional properties are, indeed, the truthmakers for these laws.

3.5 CATEGORICAL REALISM

The most widely accepted theory of dispositions is *categorical realism*.[7] Categorical realists accept the traditional analysis of dispositions in terms of subjunctive conditionals. Consequently, they hold that dispositions bear a special relationship to subjunctive conditionals – a relationship which is not possessed by other properties.[8]

The subjunctive conditionals in terms of which dispositions are traditionally defined are typically *causal* conditionals – that is, conditionals that hold in virtue of causal laws. Consequently, dispositions are generally held to be dependent on causal laws. However, on some sophisticated versions of the theory, allowance is made for the possibility that some dispositions may depend only on fundamental statistical laws.

These theories of dispositions usually depend on a basically Humean theory of causal laws. According to this theory, the strictly causal laws are all universal regularities of some sort. For reasons that are not well understood, the statements of these laws all support subjunctive or counterfactual conditionals. They tell us not only how things of various kinds actually do behave in some given circumstances, but also how they would behave in any of a range of other possible circumstances. That is, they tell us how things of various kinds are universally disposed to behave, depending on the circumstances.

The identities of the things involved in such regularities do not depend on how they are disposed to behave. For, according to the categorical realists, their identities depend only on their categorical properties or structures, which would be the same however they might be disposed to behave. Consequently, the very same things that are disposed to behave in one kind of way in this world might well be disposed to behave in very different kinds of ways in other worlds. The categorical realists thus hold that the laws of nature, and hence the causal laws on which dispositions depend, are all contingent. We call this the Contingency Thesis.[9]

It is an immediate consequence of the Contingency Thesis, and the categorical realist's theory of dispositions, that if something has a certain disposition, then, necessarily, there is a possible world in which it (or its counterpart) does not have this disposition. What is brittle here might well not be brittle there. Hence, the identity of a thing cannot depend on its dispositions; it can only depend on its genuine properties, all of which are thought to be non-dispositional. The dispositions of things cannot be (dispositional) properties; all dispositions depend on what the laws of nature are.

A metaphysical wedge is thus driven between the dispositions of things and the properties of the entities that possess them. Given that the laws of nature are contingent, the relationship between a given disposition and the categorical properties that are supposed to ground it must also be contingent, and hence the grounding properties and the disposition must be ontologically distinct from each other.[10] If this is right, then we are free to associate dispositions with categorical bases according to how the laws are in each possible world, thus ensuring that objects that are disposed to behave in a particular way in a given world are said to have the dispositions that correctly describe their behavior.

The main arguments in favor of the categorical realist's claim that dispositions need categorical bases are that they are needed to explain the continuing existence of, and also the differences between, dispositions that are not currently (and perhaps never have been, and never will be) manifested.[11] Call these the Continuing Existence[12] and the Difference Argument.

We remain unconvinced by these arguments. It is true that dispositions need to be based in reality. There must at least be properties of real things that ground them. Moreover, it is often the case that things have the dispositions they do because of their internal structures and constitutions, and in all such cases we may say that the dispositions are grounded in these structures and constitutions. However, it is not clear that the basis of any given disposition must, or must ultimately, be non-dispositional or categorical. For, without begging the question against non-Humeans, it cannot be assumed that the basis of a disposition does not, or does not ultimately, depend on genuine dispositional properties. For example, the dispositions of an object might well depend on the causal powers of its parts, as well as on how these parts are arranged.

Consider first the Continuing Existence Argument for categorical realism. A categorical basis for a disposition is needed, it is argued, because a disposition must be capable of continuing to exist unmanifested. The kinds of structural properties on which dispositions are likely to depend are certainly capable of enduring. Therefore, if a particular disposition has a basis in such properties, its continued existence is explained. It continues to exist because the properties that ground the disposition continue to exist. However, this argument does not establish the need for categorical bases for dispositions, unless it is assumed that the only properties that are capable of enduring without support are categorical. It therefore begs the question against those who think that there are ontologically basic dispositional properties that endure and support other properties. We see no good reason for making this assumption. On the contrary, we think there are good reasons for supposing the assumption to be false. Dispositional properties are capable of enduring whether or not they have categorical bases.

First, there is the Argument from Science. Many of the most basic properties are evidently both occurrent and dispositional, as I.J. Thompson

114

(1988) has argued. If such properties have categorical bases, then they are unknown to us. Moreover, there is no prospect in modern physics of being able to characterize the most fundamental existents (for example, the particles and fields of modern physical theory) by their categorical properties alone, as Prior's (1985) account of dispositions would seem to require. In fact, it is doubtful whether these things have any categorical properties at all. On the contrary, the properties of the most fundamental things in nature, including mass, charge, spin, and the like, would all appear to be dispositional.

Second, there is the Argument from the Nature of Laws. The laws of nature do not merely describe the behavioral regularities of things characterized by their categorical properties alone. On the contrary, the laws of nature describe the ways in which things are intrinsically disposed to be or behave.[13] Dispositional concepts thus occur essentially in laws. Therefore, laws of nature of the sort that Prior needs to effect her ontological reductions of dispositional properties simply do not exist. There are no regularities of behavior that are specific to things of a given shape or size, for example, or to the members of the extensions of any other categorical property, as there are laws governing the interactions of charged or massive particles.

In chemistry, there are laws that plausibly describe how substances of various kinds interact. But these laws do not just express regularities. On the contrary, they make use of precisely the kinds of dispositional concepts that Prior's theory is intended to reduce. That salt dissolves in water, for example, or that hydrogen is exploded by a spark in oxygen to form water, are laws that, perhaps more plausibly than most, are just statements of regularities. But the laws that underlie these regularities are dispositional, for what has to be explained in these cases is the solubility of salt in water, or the potential for hydrogen and oxygen to combine explosively to form water.

Third, there is the Ontological Regress Argument. Whenever a causal power is seen to depend on other properties, these other properties must always include causal powers. For the causal powers of things cannot be explained, except with reference to things that themselves have causal powers. Structures are not causal powers, so no causal powers can be explained just by reference to structures. For example, the existence of planes in a crystal structure does not by itself explain its brittleness, unless these planes are cleavage planes – that is, regions of structural weakness along which the crystal is disposed to crack. But the property of having such a structural weakness is a dispositional property that depends on the fact that the bonding forces between the crystal faces at this plane are less than those

115

that act elsewhere to hold the crystal together. Therefore the dispositional property of brittleness in a crystal depends not only on the crystal's structure, but also on the cohesive powers of its atomic or molecular constituents. However, cohesive powers are causal powers. They are the forces that bind things together. For a crystal structure, these forces are presumably electromagnetic, and therefore depend on the dispositions of charged particles to interact with each other in the sorts of circumstances that exist inside a crystal. To explain the distribution of the cohesive forces existing in such circumstances, the structure of the crystal must be described in some detail. But this description will not by itself do anything to explain the cohesion of the parts of the crystal. To do this, it is also necessary to say what energy states are occupied by the structure's various constituents, and to specify their dispositions to resist being prised out of their respective positions. So cohesive powers have to be explained in terms of other causal powers. And there never seems to be any point at which causal powers can just drop out of the account.

The Difference Argument for categorical realism is based on the intuitive belief that two substances (in the same world) cannot differ only in respect of their dispositions, or dispositional properties, if indeed there are such things.[14] For if they did differ in only these ways, the differences between them would be inexplicable. Consequently, it is argued, if two things have different dispositional properties, they must also differ in some respect that is not dispositional.

It is plausible to suppose that if two things differ in respect of any one of their dispositional properties, then they must differ in other ways as well. For example, if two people differ in mathematical ability, then we may reasonably think that they will also differ in other ways – for example, in their capacities for pattern recognition. Otherwise, the difference between them would be inexplicable. One difference can only be explained by means of another difference. But the explaining differences need not be differences of internal structure, as the categorical realist supposes. One difference of capacity can, and often does, explain another, as the explanation of brittleness shows. Moreover, the entities we are dealing with simply may not have any internal structure. They may be things belonging to fundamental natural kinds which are simple and unstructured. Such things can differ from each other only in respect of their causal powers, capacities, and propensities.

Neither the argument from Continuing Existence nor the argument from Difference proves the need for there to be categorical properties to ground dispositions; at most they establish that distinct dispositions must either be, or be grounded in, distinct occurrent properties. This is not to

116

say that dispositions cannot be grounded in properties that include categorical properties – just that they need not be. The dispositions might, for example, be fundamental properties, or properties that are grounded in occurrent properties that are themselves dispositional. For an occurrent property is not necessarily a categorical property; it may be a property that has causal potency. There is nothing in either the Continuing Existence or the Difference Arguments to prohibit occurrent properties from being causally efficacious.

The main arguments for categorical realism are thus inconclusive. Nevertheless, belief in categorical realism dies hard. Categorical realists, who mostly hold a Humean theory of laws, believe that they must posit categorical bases to ground dispositions. A dispositional base would itself require a base. The argument stems from the Contingency Thesis, which the Humean theory entails. If the laws of nature are contingent, it is argued, they can only be contingently connected with the entities they govern, and hence they must be ontologically distinct from them. And since the behavior of the entities is determined by the laws of nature, the entities in themselves must be causally impotent. This is why the Humean must regard all genuine occurrent properties as categorical. However, we reject the Contingency Thesis about laws. For we believe that there are real dispositional properties, and that the identities of the most fundamental kinds of things in nature depend on them. What makes something a neutron, for example, is its causal powers. It is not something that has an identity independently of these. On the contrary, what a neutron is disposed to do – for example, how it is disposed to interact with fields and with other particles – is what makes it the kind of thing it is. A particle is a neutron if and only if it is disposed to behave as a neutron does. Its dispositional properties are of its essence. Consequently, the dispositional properties of the most fundamental kinds of things cannot vary from world to world, as the categorical realist supposes.

We also reject the standard Humean semantics of dispositional terms. In Humean analyses, dispositional terms are defined operationally by one or more subjunctive conditionals. For example, a thing x is said to have the disposition to E in circumstances C iff x were to be in circumstances C, then x would E. On our analysis, however, dispositional terms may name occurrent dispositional properties – that is, properties whose natures are to dispose their bearers to behave in certain ways in certain circumstances. Consequently, to say that x has a disposition to E in circumstances C is to postulate the existence of an occurrent dispositional property in virtue of which it is true that x would (normally) E in circumstances C.[15]

Despite its failings, categorical realism does seem to have some attractive features. For instance, it clearly preserves the important distinction between dispositional and non-dispositional properties. For there does indeed seem to be a fundamental difference in kind between a property such as triangularity, which is not dispositional, and a property such as fragility, which is undoubtedly dispositional. We agree with the categorical realists on this point, and accept the idea that any decent theory of dispositions should preserve this fundamental distinction. We do not accept, however, that the fact that it is sometimes difficult to assign a single occurrent property to a disposition, as it is in the case of fragility, shows that dispositional properties are generally not real properties. The difficulty that arises with a disposition such as fragility is that the members of the class of fragile things are not intrinsically similar to each other, as the instances of any natural property must be. The similarities between fragile things are extrinsic and behavioral, and things are classified as fragile for practical reasons that have little to do with the intrinsic properties of the objects concerned. Vases, ecosystems, personalities, and fabrics can all be fragile, but not in virtue of any intrinsic properties they have in common. They are classified together only because all of these things need to be handled with care, if one doesn't want them to be broken or damaged.

Diverse properties can be grouped together and labeled as properties of the same kind for many reasons — reasons that have nothing to do with the intrinsic similarities of the objects that possess them. "Expensive," "convenient," "functional," "complex," and "delicious," for example, are all terms that might loosely be said to name properties. But they do not name real or natural properties — properties that can be supposed to exist in nature independently of human interests or purposes. Consequently, these "properties" should not be taken as examples of properties for the purposes of ontology. Such "properties" are always grounded in other occurrent properties, and for the purposes of ontology, it is clearly important to distinguish between these so-called "properties" and their occurrent bases.[16] Categorical realists quite rightly do this — and so do we. However, categorical realists make a serious mistake if they suppose that all dispositions are grounded in such superficial similarities as the ones just listed.

3.6 DISPOSITIONS AND CATEGORICAL REALISM

Bigelow (1999) argues that Caroline Lierse and I misrepresent the position of categorical realists on dispositions, and claims that we exaggerate the distance between their theory of dispositions and ours. The categorical realist's position that we had in mind when writing the paper to which Bigelow was responding was actually that held by David Armstrong (in Armstrong, 1978). However, as Bigelow says, it is arguable that the position taken later by Elizabeth Prior, Robert Pargetter, and Frank Jackson (in their 1982 paper, hereafter PPJ) is not as far removed from our 1994 position as Armstrong's earlier one. We agree with this. Nevertheless, there is still a considerable distance between PPJ and us, and they are still basically in the categorical realist's camp.

PPJ argue that every behavioral disposition must have a *causal basis*. That is, they hold that there must be an intrinsic property or structure in virtue of which a given thing has the behavioral disposition in question. They are non-committal, however, about nature of the causal basis, and they explicitly deny that it must be categorical (PPJ, 253). We have no reason to disagree with them so far about any of this. For it is no part of our theory that behavioral dispositions do, or could, just hang in the air, without being rooted in occurrent properties or structures. However, some (but not all) behavioral dispositions are grounded in occurrent properties of the kind we call "dispositional properties" – properties whose identities depend on the behavioral dispositions they support.

Dispositional properties support behavioral dispositions in the following sense: If an object x has a determinate dispositional property P, and x exists in any of the circumstances C_i that belong to its triggering range C, then it must, in virtue of having P, be intrinsically disposed to behave in the manner E_i, depending on C_i. Of course, an object x having P in the circumstances C_i need not actually behave in the manner E_i. Other forces could come into play which would frustrate the disposition. But if there are no other such forces, then the behavioral disposition to E_i in circumstances C_i must be displayed. Thus, on our view, the disposition is displayed because the object has the dispositional property, and exists in circumstances apt for its display. It does not have the dispositional property because the behavior E_i happens regularly to be displayed in circumstances C_i.[17] On the other hand, a behavioral disposition such as this need not, in our view, always have some other (non-dispositional) kind of causal basis.

Dispositional properties, like all other genuine occurrent properties, must be the same in all possible worlds in which they exist. Fragility, which is assumed uncritically by PPJ to represent all behavioral dispositions, is not even a genuine property. There is a predicate "is fragile" that is true of some things. But there is no unique property in virtue of which things are fragile. Certainly there are occurrent properties and structures – indeed a great many of them – that make for fragility. But the fragile/non-fragile

119

distinction is a social construct, useful for the purposes of warning people to take care, but of no use whatever in scientific explanation. There simply is no natural property of fragility. Long-stemmed glasses, ancient parchments, ecosystems, and spider webs are all in their own ways fragile, but they have nothing intrinsic or structural in common to which the name "fragility" could be attached. These things are grouped together just because they can all be broken or damaged easily.

Nor is fragility a typical behavioral disposition, for it is quite unlike any of the thousands of behavioral dispositions investigated in the natural sciences that are grounded in occurrent dispositional properties. The causal powers, capacities, and propensities that feature in scientific explanations, for example, are all dispositional properties, in this sense. For each is supportive of a wide range of specific behavioral dispositions, often an infinite number of them, and for each of them it is true that their identities depend on the sets of behavioral dispositions they support. The refractivity of a certain kind of glass, for example, is a dispositional property of the glass that grounds infinitely many behavioral dispositions. The dispositions it grounds are characterized by a range of quantitatively different circumstances (different angles of incidence, different frequencies of incident light, different refractivity of the medium, and so on) resulting in a range of quantitatively different effects (different angles of refraction, total internal reflection, and so on). This is the sort of thing we have in mind when we speak of a dispositional property. It is a causal power, capacity, or propensity that underlies a range of behavioral dispositions.

Dispositional properties such as these are clearly not categorical properties. For they could not survive changes to the laws of refraction or to the laws of quantum mechanics from which they derive. However, PPJ claim not to be committed to the view that all behavioral dispositions must have categorical properties or structures as their causal bases. Is there then any reason why they should not just accept our analysis and agree that the causal basis of a behavioral disposition might simply be a causal power? We think there is. The principal difficulty for PPJ lies in their acceptance of the Distinctness Thesis. For this would commit them to saying that in another possible world, the same causal power might ground a different set of behavioral dispositions, or conversely that the same set of behavioral dispositions might be grounded in a different causal power. For, they insist, the connections between behavioral dispositions and their causal bases are contingent. They are contingent, they suppose, because the laws of nature are contingent. To accept the notion that there are necessary connections between behavioral dispositions and their causal bases is precisely what they, as Humeans, cannot do. Therefore, despite their reticence about saying that the causal bases of behavioral dispositions must be categorical, they are nevertheless committed to the view that they are. For, given PPJ's Distinctness Thesis, the causal basis of a disposition must be able to survive changes in the relevant laws of nature.

In reply, PPJ might be willing to accept the argument so far, but argue that our so-called dispositional properties are not the real underlying causal bases for the behavioral dispositions they ground. For refractivity is itself a (complex) disposition that plausibly requires a causal basis. What then might PPJ have in mind as the causal basis of a disposition such as refractivity? We speculate that they would say that the causal basis must lie in the molecular structure of the glass, a structure that, they might well say, can be characterized without reference to the laws of refractivity or of quantum mechanics. However, what one might say by way of characterization is one thing; what the reality is, is another. For the supposed causal basis of refractivity is not categorical in the sense that it could survive even the slightest changes to the laws of nature relevant to refractivity. In reality, the molecular structure, indeed the molecules themselves, depend ontologically on these laws. So the molecular structure that is undoubtedly the causal basis of refractivity just could not exist at all in a world in which the relevant laws of nature were different.

3.7 MELLOR'S DISPOSITIONAL FOUNDATIONALISM

One philosopher who denies the reality of the distinction between categorical and dispositional properties is Mellor (1974, 1982). Mellor argues that all genuine properties are really dispositional. He does so on the ground that the only plausible way of making the distinction is with reference to their support, or lack of support, for subjunctive conditionals. Mellor (1982) claims that all physical properties support subjunctive conditionals, and are therefore really dispositional. The so-called categorical properties are no exception. If there are any physical properties that are straightforwardly non-dispositional, and hence categorical in the required sense, then they must surely be the properties of shape – for example, the property of being triangular. But, as Mellor argues, even this property supports subjunctive conditionals, and so fails the test of categoricity.

Mellor argues that because there is no proper basis for a distinction between dispositional and non-dispositional properties, the so-called dispositional properties should not be discriminated against. They have as much right to be considered genuinely occurrent as any other properties.

Mellor's second argument in favor of non-discrimination is an independent argument for the reality of dispositional properties. He cites two tests for the claim:

1. The principle of multiple manifestation – that a real property must manifest itself in more than one way.
2. Schlesinger's principle of connectivity[18] – that any real property must be nomically connected with other properties, so that two physical systems cannot differ only in respect of a single property.

Mellor claims that many dispositional properties clearly satisfy both of these criteria. Hence, these properties should be accepted as genuine ones.

Mellor's criticism of the categorical/dispositional distinction is provocative, as it threatens the very foundations of the categorical realists' ontology. The principal point his analysis illuminates is that all physical properties can be characterized operationally in terms of subjunctive conditionals, and as dispositional properties are traditionally explicated in terms of operational definitions, it follows that all properties have dispositional aspects.[19] The upshot of this argument is that the dispositional/categorical distinction so construed cuts no philosophical ice. When this fact is combined with his second point – that dispositional properties clearly have some claim to be genuine properties – Mellor seems to have mounted a persuasive argument for both reevaluating the lowly status of dispositions and questioning the role of (and perhaps dispensing with) categorical properties.

Mellor's attack on the reality of non-dispositional properties is interesting. But it must be remembered that Mellor's theory purports to be Humean. So how does his view on dispositions lend itself to a Humean ontology? At first, it seems that by denying the existence of non-dispositional properties, Mellor has eliminated occurrent properties from his ontology. But on closer inspection, it can be seen that this is not the case. His claim that dispositional properties are "real" properties suggests that Mellor is in some sense rejecting the traditional operationalist semantics of dispositional terms, and regarding dispositions as occurrent properties. In fact, his analysis seems to suggest that all occurrent properties are really dispositional. Hence his objection to categorical properties does not amount to a rejection of occurrent properties. However, there is a difficulty in reconciling Mellor's analysis with a Humean ontology. By embracing the Contingency Thesis, as he does, Mellor has to subscribe to the view that the laws that hold in this world are only contingently related to the entities that exist in it. Therefore, the behavior a thing displays must be logically distinct from the kind of thing it is.[20] But if Mellor wants to deny the existence of categorical properties, avoid embracing a behaviorist theory of dispositions, and yet have an ontology of occurrent properties, then his only option would appear to be to identify dispositions with occurrent properties that are not dispositional. But what could these be? Could they perhaps be those structural or other properties that must exist if dispositions are to exist? No, because such an identification would be inconsistent.

The identification is not permissible in a Humean ontology, for it is not consistent with the Contingency Thesis. It may be acceptable to equate a disposition with its base in this world (for example, to identify solubility with some physically specifiable structural property of molecules). But such an action only gains its legitimacy by the fact that the laws in this world are fixed. In another world, where the laws are different, a Humean must hold that the supposed identity relation might not hold. Consequently, any identity relation between a disposition and its base, and hence between an entity and its behavior, would at best be a contingent relation. This seems to conflict with Mellor's analysis.

Perhaps his theory could be rescued by embracing a new semantics of dispositional terms. However, we doubt if this would remove the problem of the logical connection between properties and the events they manifest. It seems that Mellor must either (1) abandon the Contingency Thesis, (2) reinstate categorical properties, or (3) provide a new semantics of dispositional terms that can explain their occurrent nature in a Humean ontology. In the next section, we advocate doing all three of these things, but, as will be evident, our analysis of dispositional properties has a distinctly anti-Humean flavor.

3.8 AN ANALYSIS OF DISPOSITIONS

Dispositional properties are attributed to things to explain the dispositions they manifest – that is, to explain how things will or be likely to behave in various kinds of circumstances. Such explanations are easily parodied. For they often appear to be trivial. The manifest disposition of takers of a given drug to go to sleep following its ingestion is only trivially explained by saying that the drug is a soporific. Nevertheless, this is an explanation, and it is not the only possible one. The drug taker might believe the drug to be a soporific, when in fact it is only a placebo, and the disposition to sleep might well be caused by this belief, rather than by the nature of the drug that is taken.

The dispositional properties of things are the causal powers, capacities, and so on that are postulated to explain the dispositions of things. In the simplest kind of case, the dispositional property is one that is linked essentially and directly to a certain natural kind of causal process – the kind of causal process that is essentially a display of this property. Of course, a causal process that is superficially like the processes of a given natural kind might be faked, or be due to some combination of other dispositional properties (as the case of the placebo soporific illustrates). So genuine dispositional properties cannot be defined behavioristically. They are postulated to explain what is taken to be a natural kind of process, and natural kinds of processes, like natural kinds of objects, are designated, rather than defined. Nevertheless, it may be possible to identity a dispositional property descriptively. For it may be sufficient to secure reference to the kind to specify in behavioristic terms the kind of process that is typical of it. The identification of natural kinds of processes is not fundamentally different from the identification of natural kinds of substances. Descriptions have a role in fixing reference to a kind, even though the real essence of the kind is something different.

A natural kind of process that is a display of a dispositional property has a certain real definition. And it is one of the primary objects of science to try to discover what the real definitions of the various natural kinds of processes are. In the case of any simple causal process, the real essence will be a dispositional property, and the scientific problem will be to specify precisely what this property is. In general, the real essence of a causal process of a given natural kind will be specifiable counter-factually by the kind (or kinds) of circumstance C in which it would be triggered, and the kind (or kinds) of outcome E which would (or would with probability p) result, if there were no interfering or distorting influences. In the simplest kind of case, in which the dispositional property can be triggered or displayed in only one way, the dispositional property may be uniquely characterized by an ordered pair $\langle C,E \rangle$, where "C" denotes a kind of circumstance and "E" a kind of event. If x is an object that has this dispositional property, then x may be said to have the power, capacity, or propensity to E in circumstances C. However, it is not an *a priori* matter what the real essences of the natural kinds of processes are, and what is being determined is not the meaning of a dispositional term.

Real dispositional properties ground natural kinds of processes, and like all natural kinds, exist independently of our systems of classification. Natural processes that appear to be of the same kind may turn out to be essentially different, and processes that appear to be very different may belong to different species of the same kind. Refraction through a prism and diffraction from a grating produce very similar outcomes. Nevertheless, they are essentially different kinds of processes. One results from the refractivity of a medium, the other does not. On the other hand, many of the most important discoveries in science result from identifying apparently very different kinds of processes as species of the same generic kind. Newton, for example, showed that the apparently different kinds of processes of falling toward the earth and orbiting the sun are essentially the same. Similarly, Lavoisier showed that respiring, rusting, and burning are all essentially processes of oxidation. Malcolm Forster (1988) talks of discovering a common cause in these and similar cases. We think that these discoveries are best described as discoveries of sameness of essential nature.

We suppose the natural dispositions to be simply the real essences of the natural kinds of processes they ground. That is, we suppose that an object cannot participate in a process of a given natural kind – for example, to E in circumstances C – unless it has the requisite dispositional property of E-ing in these circumstances. In contrast, artificial or socially constructed kinds of processes do not have real essences, and the dispositions defined with reference to such processes are not grounded in corresponding dispositional properties.

Natural kinds of processes may be either causal or stochastic. An example of a natural kind of stochastic process is that of β-decay. It is essentially the spontaneous emission of an electron from the nucleus of an atom, resulting in an increase of its atomic number by one. This process occurs independently of human concerns and has its own essential nature. To specify a kind of stochastic process, it is necessary and sufficient to say what kind of transition it concerns, and how probable it is that such a transition will occur within any given time interval, in the absence of other causal influences.

The dispositions of which we speak may be distinguished as "real" or "mere Cambridge" dispositions, depending on whether or not their displays involve natural kinds of processes. The displays of real dispositions are natural causal processes emanating from the objects that possess them, and involve real changes in the objects taking part in them. Solubility, for example, is a real disposition, for the process of solution is a natural kind of process, and a soluble substance undergoes a genuine change when this disposition is manifested. Triangularity, on the other hand, is not a real disposition,[21] although it might be said that a particular object has the disposition to look triangular, or be such that if you were to count its corners correctly, you would get three. For being looked at and being counted are

125

not natural kinds of processes emanating from any triangular objects that may be subject to this treatment, and their outcomes involve what Geach referred to as "mere Cambridge" changes[22] in the objects being looked at or counted. Hence they are, at best, mere Cambridge dispositions.

Although we reject Mellor's analysis of dispositions, which allows him to count triangularity as a genuine dispositional property, we agree with him in accepting Schlesinger's connectivity criterion for the reality of properties. The principle of multiple manifestation, which he also cites, is an appropriate criterion for the reality of entities postulated as the bearers of properties, but it is not so obvious that properties must also manifest themselves in more than one way, as this principle requires. We hold that a difference in respect of any real property must make a difference – that is, have some effect. But to have an effect, it is sufficient if it is nomically connected with other properties. Hence, this requirement is satisfied if the principle of connectivity is satisfied; hence the importance of this latter principle.

Most paradigmatically categorical properties clearly pass the test of connectivity. Differences in shape, size, and other categorical properties make a difference because different spatio-temporal relations, and hence different structures, make a difference. But just as categorical properties pass the test of connectivity, the same is true of many dispositional properties. Inertial mass, for example, is nomically connected with other properties. Therefore, by the test of connectivity, there is at least as much reason to count the inertial mass of an object as a genuine property as there is to consider its shape to be a real property.

The main difference between the analysis of dispositional properties we are proposing and its more traditional rivals lies in the semantics of dispositional terms used to refer to them. We reject the operationalist way of defining dispositional *terms*. We allow that scientific investigation may lead to specific characterizations of dispositional *properties*. But these characterizations are not verbal definitions. They are proposed as real definitions of real properties. In our theory, when we seek to discover the essences of dispositional properties, we are not seeking to define the terms we use to refer to them. We *denote* them and then seek to explicate them. And because we are dealing with natural kinds of properties and processes, we can be wrong about them. For real dispositional properties exist as distinct entities, prior to any nominalist or operationalist definitions of the terms we might use to refer to them.

We believe that this analysis of dispositions has some distinct advantages over more traditional theories. First, it explains why dispositions bear special relationships to subjunctive conditionals. Dispositional properties support subjunctives because their existence entails that certain kinds of natural processes would occur in certain kinds of (possibly idealized) circumstances to the objects that have these properties. The subjunctive conditionals simply spell out these implications. Second, it explains why dispositional properties can be mocked or frustrated. For circumstances can often be manipulated to make an object appear to have a dispositional property it does not have, or appear not to have a dispositional property it does have. Third, it explains why genuine dispositional properties can often be obscured. They can be obscured because different processes can occur in the same thing at the same time, so that the effect of any single dispositional property being triggered may well be obscured by the effects of other dispositional properties that are being simultaneously manifested.

Our analysis is "bottom up" in that for every dispositional property, there is a particular subjunctive (or set of subjunctive conditionals) that the dispositional property supports.[23] However, the converse is not true; a true subjunctive conditional does not entail the existence of a corresponding dispositional property. For example, to borrow an example from Mellor, the subjunctive conditional "if you were to count the corners correctly, you would get three" does not, on our analysis, name a dispositional property. An attractive feature of our analysis is that it leaves dispositional properties to be identified and explicated rather than defined operationally. And the process of explication is not philosophic, linguistic, or lexicographic. It is *a posteriori* and almost paradigmatically scientific.

3.9 DISPOSITIONAL ESSENTIALISM

The position we wish to defend, which we call "dispositional essentialism," is a species of dispositional realism. It is realist about the dispositional properties of the fundamental particles and fields, for example, and it is essentialist for two reasons: First, because it holds that these properties are among the essential properties of these particles and fields, and second, because it holds that it is essential to the natural processes in which these particles and fields may be involved that they be displays of these dispositional properties. We do not claim, as some philosophers have, that these fundamental dispositional properties are the ontological basis of all properties. On the contrary, we believe that there are equally fundamental categorical properties – for example, spatio-temporal relations and structures. We see no reason to suppose that such properties can be ontologically reduced to dispositional ones.

127

Real dispositional properties, we hold, may supervene on categorical properties, but never on categorical properties alone. If a dispositional property supervenes on other properties, then the subvenient class must include at least one property that is itself a dispositional property. One disposition may be ontologically dependent on another disposition, just as one causal process may depend ontologically on another. But a dispositional property cannot be ontologically dependent only on what is not dispositional. A causal power is more than just a constant conjunction.

Categorical realists seek to deal with this difficulty by claiming that dispositional properties supervene, not only on categorical properties, but also on the laws of nature. However, the laws of nature are not the right category for the ontological reduction of properties. Laws are not things that exist in the world; they are things that are true of the world. The truth-makers for the laws of nature might well be things on which dispositional properties could depend ontologically.[24] Indeed, if we are right, then this is so. For the truth-makers for the relevant laws of nature are, we hold, just the fundamental dispositional properties. Thus, we argue that while real dispositional properties may well be supervenient on other properties, this is possible only if the subvenient class includes at least some real dispositional properties. Consequently, if there is to be no infinite regress of ontological dependence among properties,[25] there must be some ontologically basic dispositional properties. Most plausibly, these ontologically basic properties are just the causal powers, capacities, and propensities of the fundamental natural kinds.

Such properties are evidently dispositional. A causal power is a disposition of something to produce forces of a certain kind. Gravitational mass, for example, is a causal power: It is the power of a body to act on other bodies gravitationally. A capacity is a disposition of a kind distinguishable by the kind of consequent event it is able to produce. Thus, for x to have the capacity to do Y is for x to have a disposition to do Y in some possible circumstances. Inertial mass, for example, is a capacity. It is the capacity of a body to resist acceleration by a given force. A propensity is a disposition to act in a certain kind of way in any of a wide range of circumstances. For example, the propensity of a radium atom to decay in a certain way in a certain time is a disposition that the atom has in all circumstances. Moreover, the dispositional properties of the fundamental natural kinds would also appear to be basic in the required sense. If they are ontologically dependent on other properties, then it is hard to see what these other properties could possibly be.

Finally, these basic properties would appear to have precisely the prop-

128

erties we should require of truth-makers of the causal and statistical laws concerning the behavior of things belonging to the fundamental natural kinds. For the existence of these causal powers, capacities, and propensities is sufficient to guarantee that these laws must hold for things of these kinds. They must hold for things or these kinds because these properties are among their essential properties. Hence, things of these kinds necessarily have these dispositional properties, and are bound to behave accordingly.

How the relevant laws of nature are grounded in the essential properties of fundamental natural kinds of things can be explained by direct appeal to the nature of these properties. For what dispositional properties do is dispose the things that have them to behave in certain ways, depending on the context. What science observes and codifies are the manifestations of these dispositions. Hence, laws that describe how dispositional properties act will at the same time tell us what those things that have these properties essentially must do in virtue of being things of the kinds they are.

3.10 DISPOSITIONAL PROPERTIES AND CAUSAL LAWS

There are two broad categories of dispositional properties, *causal* and *stochastic*. Causal dispositions refer to causal processes, stochastic dispositions to stochastic processes. Where we have a causal disposition, there is typically a certain pattern of cause-and-effect or stimulus-and-response that anything having the disposition would normally display if it were appropriately caused or stimulated to do so. A *stochastic disposition,* on the other hand, is a propensity of some kind, in which the antecedent condition is not strictly the cause of its manifestation, but only a necessary condition for it. For example, the disposition of a radium atom to decay in a certain way is a stochastic disposition. If this species of radioactive decay is to occur, it is a necessary condition that radium atoms exist. But events of radioactive decay are not caused by the existence of such atoms. Nor, as far as we know, are they caused by anything else. There is just a certain objective probability p that within any given time-interval d, such an event will occur.

Causal Dispositions and Causal Laws

Let D be the causal disposition $\langle C, E \rangle$ and $D(x,t)$ the proposition that x has this disposition at t. Then $D(x,t)$ is the claim that an event or state of affairs of the kind C occurring to x at t would, or would at least be likely to, cause an event of the kind E. Let $C(x,t)$ be the proposition that an event or state of affairs of the kind C exists or occurs to x at t, and $E(x, t + \delta)$ the proposition that an event of the kind E occurs to x in the time interval from t to $t + \delta$. Then, to say that x has the causal disposition $\langle C,E \rangle$ at t is at least to say that for some δ, the probability of $E(x, t + \delta)$, given $C(x,t)$, is greater than one-half. Actually, it is to say more than this, because the claim is not only that an E-event occurring to x by $t + \delta$ is made probable by a C-event occurring to x at t, but also that it is likely to be caused by such an event.

For a given δ, the probability that an event of the kind E will occur to x by $t + \delta$ as a result of an event of the kind C occurring to x at t depends on how strongly x has the disposition $\langle C,E \rangle$ at t. If, for some finite δ, an event of the kind E must occur to x by $t + \delta$ as a result of a C-type event occurring to x at t, then the disposition $\langle C,E \rangle$ of x at t may be said to be *causally determinate*. If all instances of a given disposition must always be causally determinate, then the disposition itself may be said to be causally determinate. In that case, there is a deterministic law of action of the disposition. Such a law is clearly a *causal law*, for it is fully determinate of what the effect of a C-type event occurring to x will be.

To illustrate our concept of a causally determinate disposition, consider, once again, electric field strength. The field strength $\mathbf{E}$ at a point P in an electrostatic field is the electrostatic force per unit positive charge placed at P. Hence, if x is the field that exists at P, $C(x,t)$ is the proposition that a positive charge e^+ is placed at P at t, and $E(x,t)$ is the proposition that e^+ is consequently subject to an electrostatic force $\mathbf{E}$ at t ($\delta = 0$ in this case), then the following law must hold:

$$\forall x,t \; [C(x,t) \Rightarrow E(x,t)]$$

This law, which states the law of action of the disposition $\mathbf{E}$, is a typical causal law. $\mathbf{E}$ is what we may call a *causal power*. It is a property of the field at P.

Stochastic Dispositions and Statistical Laws

There are at least two kinds of reasons why a disposition may not be causally determinate. First, some dispositions suffer from incurable vagueness. To define fragility, for example, we cannot do much better than say that a fragile object is one that is likely to break if dropped or otherwise

130

handled roughly. Any more precise definition might capture some more specific concept of fragility. But it would not be the broad but vague concept with which we are familiar. Dispositions, such as fragility, which, because of their vagueness, are not causally determinate could obviously exist in a deterministic world. For the kind of vagueness that attaches to such dispositions is a function of language, not of reality. The indeterminacy of such dispositions is due to their lack of specificity, not to how the world is, and in a deterministic world, they should at least in principle be eliminatable in favor of causally determinate dispositions. Indeed, in a deterministic world, all dispositions that for any reason are not causally determinate must be ontologically dependent on dispositions that are causally determinate. This is what it is for the world to be deterministic.

In an indeterministic world, such as ours, there must be dispositions that are not causally determinate for a different reason. They are causally indeterminate, not because of vagueness, but because of the indeterminacy of the underlying physical processes. Such dispositions might also be imprecisely defined, but they are indeterminate for another reason as well. For example, the probability that a radium atom existing at t will have decayed by $t + \delta$ is, for any given frame of reference, a precisely specifiable function of δ, and this probability is independent of the circumstances in which the radium atom exists. Hence, we cannot even in principle eliminate this causally indeterminate disposition in favor of any more precisely defined dispositions that are causally determinate.

Causally indeterminate dispositions such as these are *propensities,* and their laws of action are statistical laws. The statistical law follows from the fact that if anything x has a propensity $\langle C,E \rangle$ at t, then for any given value of δ, there must be an objective probability $p(x,\delta)$ that if x were to exist in circumstances of the kind C at t, then an E-type event would occur to x by $t + \delta$. This is what we call the law of action of the propensity. Things having this propensity must behave according to this law.

Statistical laws of interaction between things are often much more complicated than the simple laws of radioactive decay. In general, we have to deal, not only with the causal powers and propensities of things taken individually, but also with their responsiveness to each other, and to the various forces they generate. Therefore we need to know about the capacities of things to interact, and the probabilities of various interactions' occurring. We assume, without arguing the case here, that all such knowledge is ultimately knowledge of dispositions.

From what has already been said, it is clear that causal dispositions can sometimes ground causal laws and that stochastic dispositions can sometimes ground statistical laws. We have not shown that all causal and statistical laws can be similarly grounded, but it seems to us to be very probable that this is so. To suppose otherwise is to suppose that other causal or statistical laws have different ontological foundations, and we know of no good reason to think that this might be so.

3.11 MEINONGIANISM

In general, we can explain what we think we know if we can show how it fits in with other things. For explanation proceeds by epistemic location – by embedding what we think we know in some wider context of knowledge. Something like this is the concept of explanation required for metaphysics. In ontology, the problem is to explain the reasonableness of believing in various kinds of postulated entities. To do this, it is necessary to show how entities of these kinds can serve as reasonable and adequate bases for explaining the existence of other things. To be reasonable and adequate, the proposed basis cannot be just an ad hoc construction. An acceptable explanation must have some unifying power.

Causal powers of the kinds involved in scientific explanations have the required unifying power. For such explanations are generally quantitative, and the quantitative character of explanations in terms of causal powers makes it possible for one causal power to explain an infinite range of quantitatively distinct dispositions. For example, the disposition to E_1 in circumstances C_1 is different from the disposition to E_2 in circumstances C_2, even if the differences between E_1 and E_2 and between C_1 and C_2 are just quantitative. But the same causal power might well be responsible for a whole spectrum of such specific quantitatively distinct dispositions. The causal power that is responsible for an acceleration a_1 of an object of mass m_1 at a distance d_1 from the body in which this power resides might well be the very same as that which is, or would be, responsible for an acceleration a_2 of an object of mass m_2 at a distance d_2. Thus, causal powers may (and usually do) have quantitatively different manifestations, depending quantitatively on the circumstances.

Causal powers not only have quantitatively different manifestations, depending on circumstances, or on the quantitative properties of the objects on which they act, they may also differ from each other quantitatively, and so act with greater or less force on the same or identical things. Additionally, causal powers are often directional. For example, they may be centripetal, centrifugal, or circulatory. All of this complexity adds to the fruitfulness of causal power explanations and to their embeddedness in the broad context of scientific knowledge. Consequently, explanations given in terms of causal powers can succeed, as perhaps no other kinds of explanations can, in locating epistemically what is to be explained in a broad network of causes and effects.

While acknowledging that causal power explanations may have these desirable features, David Armstrong (1999a) argues that the account given here of causal powers and other dispositional properties implies that they are Meinongian entities.[26] If they are, then this is certainly a difficulty for the theory. Armstrong's argument is this: Generally speaking, things do not manifest their causal powers at all times. Indeed, some things may have causal powers that they never manifest. A causal power that had no categorical basis, and was never manifested, would have to be purely dispositional. One could not even give a relational account of it. To believe in such a property is therefore to believe in a property the essence of which involves a relation to an event that never occurs. It "points" to its manifestation, but "the manifestation does not occur." Such a property is a Meinongian property, Armstrong says, since its existence entails a relation to a non-existent event.

My response is this: A causal power is a generic potentiality – a property that can manifest itself in many different ways, depending on its circumstances. So it is not a property that "points" to its manifestation, as though there were just one such event. It points, if anything, to a whole range of possible manifestations. This range of possible manifestations is a set of possible events, all of which would belong to a certain natural kind if they were to occur. The infimic species of this natural kind include all of the ways in which the causal power could be manifested. In most cases, the range is infinite. Let us call this generic dynamic kind to which all of the different manifestations would belong "the effect kind, E." Now it is true that the causal power "points" in some sense to its effect kind. However, I do not see this as a problem. For the existence of the effect kind E is guaranteed if any instance of this natural kind exists (see Section 2.5) – that is, if something, somewhere, at some time, has an effect of this generic kind. However, a causal power is not just a relation of an object to an effect kind. For it also involves a relation to a set of events or circumstances apt for the production of events of this kind. These events or circumstances are also presumably events belonging to a natural kind. Let us call this other kind of event or circumstance "the causal kind, C." Then a causal power P involves a relation to two generic natural kinds of events – those of the causal kind C that determine when the power will be activated, and those of the effect kind E that result from its activation.

133

Events of these two kinds can in principle involve many different objects. But there are two species of them which are of special interest – those of the causal kind that would activate the causal power P_a of the object a, – namely C_a – and those of the effect kind that would result from its activation – namely, E_a. But this is still not the whole story. For there is a specific process set in train by whatever event of the kind C_a occurs, which results in a specific event of the kind E_a occurring. That is, there is a functional relationship between the events of the causal kind and those of the effect kind. The process that is activated by an event of the causal kind C_a is a causal process, and the functional relationship that holds between events in C_a and E_a is the law of action of the causal power P_a in the circumstances C_a.

If this is right, then a causal power is a property that is made manifest (if at all) by the occurrence of a causal process of a certain kind – one that involves this object as the bearer of the power, and one that is characterized by a law of action that is a special case of the general law of action describing causal processes of this kind. So in a sense it is true that a causal power points to a manifestation of a kind that may never occur. But manifestations of the general kind of which any given manifestation would be an instance must exist if this general kind exists. Therefore what we have is a relationship between an object (the bearer of the causal power) and a generic kind of process (the kind of causal process of which any manifestation of the power would be an instance).

Natural kinds, as I have argued (Chapter 2), are generalized universals. Classical universals are the infimic species of natural property kinds. Quantities, I argued, are spectral property kinds. As such, they normally cannot be constituted by their infimic species. That is, a quantity may exist, even though not every species of that quantity exists. The effect kinds to which objects with causal powers are related are themselves normally spectral kinds, which likewise cannot be constituted by their infimic species. Hence a given effect kind may exist even if only one effect of that generic kind exists. It does not matter that some specific kind of effect of some specific kind of stimulation of an object in some specific kind of circumstances does not exist. The existence of the relevant effect kind does not depend on such facts. For effect kinds, being quantitative, normally cannot be constituted by their infimic species.

The same is true of generic kinds of causal processes. Being quantitative, they cannot be constituted by their infimic species. Therefore the natural kind of causal process to which an object must be related by virtue of having a given causal power does not depend ontologically on the occurrence of the specific kind of process that would occur if the causal power were to be activated. It depends only on the occurrence somewhere, sometime, of a causal process of this generic kind.

Armstrong's Meinongian objection confuses kinds of events or processes with instances of those kinds. A causal power does not point to some possibly non-existent event (manifestation), although it does imply the existence of some possibly uninstantiated relations between certain kinds of circumstances in which the object may exist, and the kinds of events or processes that may consequently occur. I am therefore not much impressed with the Meinongian objection. Causal powers are not infimic dispositions, as the objection seems to presuppose, but the grounds or sources of these dispositions. Therefore it is not the case that causal powers involve any specific manifestations essentially, or point to any specific, but perhaps non-existent consequences, as the objection claims. But nor are causal powers properties that have identities that are independent of the dispositions they ground, as the categorialist maintains. For the causal powers of an object are the dispositional properties of that object that are the real essences of the natural kinds of processes (namely, causal processes) that involve that object in the role of cause. Causal powers thus involve essentially relations – causal relations – not between particular events but between certain natural kinds of events. Hume was right about this, at least. Causal relations are not primarily relations between particular events (as Armstrong believes) but relations between kinds of events.

3.12 CATEGORICAL PROPERTIES AND RELATIONS

There is another difficulty with our account of dispositional properties and of properties generally. We do not say that all properties are dispositional. We think that the intrinsic properties of the most fundamental things in nature are all dispositional. But there are many structural properties of, and relations between, things that depend on spatial, temporal, and numerical relationships that are not dispositional – properties such as shape, size, aggregation, and so on. If these structural properties and relations are not dispositional, the argument goes, then they can have no powers, and so cannot affect us. So are we not in the position we accuse Armstrong and other categorical realists of being in – of having to concede the possibility that the underlying structures that we suppose to exist in the world are unknown and unknowable?

Armstrong (1999a) raises this difficulty in discussing what he calls Swinburne's Regress. He does not think that Swinburne's Regress is much of a problem for us, but he does think that the problem of our knowledge of categorical properties and relations is a real difficulty. Swinburne (1983) argued that there must be properties and relations that are not reducible to powers. Here is his argument: All manifestations of powers are ultimately changes in the properties or relations of things. If such changes were just changes in the powers of things, then we could know about them only if they were manifested. But we could know about these manifestations only if they in turn were manifested, and so on to an infinite regress. As Armstrong notes, we need not be disturbed by this argument, because we do not deny that there are genuine categorical properties and relations. But in that case, protests Armstrong, how can we know about them at all?[27]

Our answer is that we know about them in the sort of way that we know about such properties in perception. Spatial properties, such as shape and size, are known to us because things of different shape or size affect us differentially. They produce in us different patterns of sensory stimulation, so that things of different shapes or sizes look or feel different. They also behave differently, and different patterns of behavior, such as rolling or sliding, are readily associated with different shapes. But we do not think that there must be a distinctive causal power for each different shape. The different shapes are reflected in the different patterns of sensory stimulation, and these different patterns arise, we suppose, from the different patterns of distribution of the causal powers in the world we are observing.

The manifestations of the dispositions grounded by causal powers must be either changes in the causal powers (or liabilities or propensities) of things, changes in their circumstances (for example, of their states of motion, agitation, and so on), or changes in their composition or structure. Some of these changes will result in the extinction of things or in the creation of new things (when essential properties or structures are changed). Some of these changes will be changes occurring in us, or in our measuring or recording instruments. Some of the changes will be what I should call changes of categorical properties or relations, some of which will be accidental properties of, or relations between, the things affected, and some will be essential properties. Others will be changes of causal powers, capacities, and so on, and again, some of these will be changes in the accidental properties of things and some of essential properties.

Swinburne's Regress depends on the assumption that the only changes that can occur are changes in the causal powers of things, so that causal powers can beget only new or different causal powers, which in turn can beget yet other new or different causal powers, and so on. But causal powers, to be effective, have to be activated. So, if we grant Swinburne's as-

136

sumption, the question is whether the activating of a causal power can be effected by an event that is itself just an alteration, creation, or extinguishing of a causal power. I do not see any reason for excluding this possibility, but examples of such actions do not readily spring to mind. Causal powers are generally activated by changes of circumstances, which are generally not changes in the causal powers of things. They are activated, for example, by changing the spatial locations of things, or their motions or orientations, or the surrounding fields, and so on – all of which involve changes of relations and none of which involves changes in causal powers.

The question for us is whether we can make a satisfactory distinction between (a) the causal powers and propensities of things, and (b) the relations that (typically) characterize their structures or circumstances. To describe a causal power, one has to say how a thing with this power must (in virtue of its possession of this power) act in the various circumstances in the complex range of circumstances in which it would be a causally relevant factor. That is, one must specify its laws of action. The laws of action of any causal power will always, or nearly always, be descriptions of the changes that would occur (in the ideal circumstances in which there are no other relevant causal factors) in the relations between things, or in the structures of things, in response to various possible stimuli (which themselves will normally be just other possible changes in the relations between, or structures of, things).

Now, in our view, the relations involved in causal laws are not causal powers, or propensities or liabilities, or anything of the sort. They have a different role in causal explanation, and are described differently. The relations in question are relations between things of various kinds, and such relations must be described if we wish to describe the circumstances in which causal powers may be exercised. The things related will no doubt have causal powers that will be relevant to their behavior. But neither the relations themselves, nor any changes in them, are causal powers, even though such changes are the phenomena we usually call "causes" and "effects."

It is true that knowledge of relations depends on the causal powers of the things related. But it is not true that the relations are themselves causal powers. The relation of being separated by one meter is not a causal power or propensity, although it may be a relevant factor concerning the actions of such properties. Things separated by one meter may be more attracted to each other gravitationally, or more strongly repelled electrically, than similar things separated by two meters. However, it is one thing to be a relevant causal factor, because it is involved essentially in a causal relationship; it is another to be a causal power.

137

The relations of spatial separation, for example, are independent of the natures of the things separated (once it is settled what are to count as their locations). Therefore, these same relations could exist in a world that is very different from ours – for example, in which properties such as mass, charge, spin, and the like did not exist. They could, for example, exist in a classical mechanistic world in which things had no properties at all, other than the passive primary ones. In such a world, the spatial and temporal relations between things would be unknowable. But, conceptually, it is a possible world. On the other hand, the identities of the properties we can know about, because they have the power to affect us or our measuring or observational instruments, depend on their laws of action – laws that involve various spatial and temporal relationships essentially. Therefore, these properties could not exist in worlds without spatial or temporal relationships. The spatial and temporal relationships on which all observable properties depend essentially are therefore more fundamental ontologically than any of the causal powers.

But if spatial, temporal, and other primary properties and relationships are not causal powers, the question arises as to how we can know about them. We can know about them, we say, because of the dependence of the quantitative laws of action of the causal powers on these relationships. If the laws of action of the causal powers were independent of such factors as size, shape, direction, duration, spatio-temporal separation, and the like, then we could never know about them, just as the objection to our account suggests.

NOTES

1. See Section 3.5.
2. This section and the remaining sections in this chapter are based on the joint paper I wrote with Caroline Lierse. Accordingly, I write on her behalf as well as mine, and use the first person plural.
3. See, for example, Goodman (1955, 40–1), Popper (1962, 70), and Mellor (1974).
4. For example, Harré (1970) and Harré and Madden (1973, 1975) have argued against Hume that things have causal powers that derive from their essential natures. Shoemaker (1980) has defended the claim that properties are essentially distinguished from each other by what they dispose their bearers to do; hence the dispositions of things are of the essence of the properties they possess. Fales (1986, 1990) has argued that the essential properties of the most fundamental natural kinds are their monadic properties, and insofar as these properties are dispositional, things of these kinds must behave as these properties prescribe.
5. See Fales (1990), Ch. 8, Section 2.
6. Shoemaker (1980) distinguishes between *dispositional* (flexible, soluble, malleable) and *non-dispositional* (e.g., square, round, made of copper) predicates. For he thinks

the term "dispositional," and presumably also "categorical," should be reserved for use in connection with predicates. He then goes on to make another distinction between "powers ... and the properties in virtue of which things have the powers they have" (p. 113). One can think of a power, he says, "as a function from circumstances to effects." But this is very confusing. Certainly we need a distinction between dispositional predicates and dispositional properties. For not every significant predicate denotes a property and not every property is nameable. But the genuine dispositional properties are presumably just those properties Shoemaker would call causal powers. If so, then we prefer the name "dispositional properties" for them because it is more general in its signification. It obviously includes, as "causal powers" does not, the more passive capacities and propensities of things that also ground dispositions. There is no need therefore for any further properties to ground the powers, unless one thinks, as categorical realists do, that all causal powers require categorical bases. A causal power might of course be ontologically dependent on other properties (although it need not be), but these other properties need not themselves all be causal powers.

7. For a general defense of categorical realism, see Armstrong (1968, 1973), Pargetter and Prior (1982, 1985), Prior, Pargetter, and Jackson (1982).

8. Prior (1982) is among those who argue that this is one characteristic of dispositions that distinguishes them from categorical properties. However, Mellor (1982) rejects this distinction by arguing that all paradigmatically categorical properties entail subjunctive conditionals.

9. This is not only accepted by Humeans who hold a regularity theory of laws; it is also widely accepted by many who would reject the regularity theory. It is accepted, for example, by Dretske (1977), Tooley (1977), and Armstrong (1983), who argue that the laws of nature are contingent relations between universals.

10. This is Prior, Pargetter, and Jackson's (1982) Distinctness Thesis.

11. Armstrong (1968, 85–7) argues that dispositions necessarily have bases, although in a later work (Armstrong, 1973, 13–5), he admits to the possibility of "ultimate potentialities" – that is, an endless regress of dispositions. Prior (1985) argues that it is a matter of fact that the bases of dispositions are categorical properties (p. 67).

12. The Continuing Existence argument (although not under this name) is a standard argument for categorical realism. It is used by Armstrong (1968) and Prior (1985) to establish the need for categorical bases for dispositions. Formally it is: (a) By definition, if X has the disposition Y, then the basis of Y is the set of occurrent properties of X (other than Y) in virtue of which X has Y. The premises are:

    1. Dispositions continue to exist unmanifested.
    2. The fact that dispositions continue to exist needs explanation.
    3. The continued existence of a disposition would be explained if it had a purely categorical basis, for the continued existence of such a basis needs no explanation.
    4. The continued existence of a disposition cannot be explained in any other way.

    Against this we argue:

    5. The set of occurrent properties or structures of X in virtue of which X has Y is either non-existent or not purely categorical.

6. If there is no basis for Y, then Y is a fundamental property.
7. If the basis for Y is not purely categorical, then it includes dispositions that are either fundamental properties or have bases that are not purely categorical.

Therefore,

8. The continued existence of every disposition is inexplicable unless there are some (fundamental) dispositions whose continued existence does not need to be explained.

Therefore 2 must be rejected.

13. This is explained more fully in Section 3.9.
14. The Difference Argument is this:

   1. The following propositions cannot all be true:

     (a) Y is a disposition.
     (b) A has Y but B does not have Y.
     (c) A and B are otherwise identical.

   2. In all cases where (a) and (b) hold, (c) is false.

   Therefore,

   3. In all cases where (a) and (b) hold, there is some categorical difference between A and B that explains why A has Y, but B does not have Y.

Against this, we argue that the premises do not entail the conclusion. We are inclined to accept 1 because Schlesinger's principle of connectivity – that if (a) and (b) hold, then there will be some other difference between A and B – seems plausible, but we deny that this difference needs to be, or even could be, purely categorical.

15. The analysis we are offering is thus similar to Fales (1990, Ch. 8, Section 2).
16. For a discussion of the distinction between real kinds and properties, on the one hand, and nominal kinds and properties based on superficial similarities of appearance or function on the other, see Sections 2.7 and 2.8.
17. This is my answer to Bigelow's (1999) worries stemming from the *Euthyphro*.
18. See Schlesinger (1963, Ch.3).
19. Shoemaker (1980) and Swoyer (1982) both make the point that we can only know about properties by observing their effects in various circumstances. Consequently, to know what properties a thing has, we need to know what its dispositions are. However, this conclusion does not follow, unless dispositions are interpreted as including mere Cambridge dispositions.
20. In reply to an objection by Mackie, Mellor explicitly states that logical connections are permitted between things and events. But he denies that this conflicts with Hume's principle that there can be no logical connections between distinct existences. Mellor (1974) argues that Hume's principle does not apply to a heterogeneous ontology of things and events. However, if the laws of nature are contingently related to the things that exist, then the kinds of behavior (events) manifested by these things are only *contingently* connected to them. Hence, even

if Mellor embraces a more restricted version of Hume's principle, his analysis of dispositions is still inconsistent with the Contingency Thesis.

21. Pace Mellor (1974, 1982)

22. The terms "Cambridge" and "mere Cambridge" were introduced by Geach (1969, 71) to distinguish between real and non-genuine changes. See also Shoemaker (1980). Geach's "Cambridge Criterion" for a thing having changed is as follows: "The thing called 'x' has changed if we have '$F(x)$' at some time t true, and '$F(x)$' at some time t' false, for some interpretation of 'F', 't' and 't''." Geach acknowledges that this definition is intuitively unsatisfactory for it includes cases of change that we would not wish to count as instances of genuine change. For instance, according to the Cambridge Criterion, Socrates would change by coming to be shorter than Theaetetus in virtue of the latter's growth. Geach refers to this kind of change as a "mere" Cambridge change. It should be noticed that Cambridge changes in general include "mere Cambridge" changes.

23. Similar "bottom up" analyses of dispositions are to be found in Harré (1970), Harré and Madden (1973, 1975), Shoemaker (1980), Swoyer (1982), and Fales (1990). These theories are similar insofar as they all propose that objects have powers or essential natures whose existence entails the manifestation of the disposition when the appropriate conditions are realized.

24. For an analysis of the concept of truth-maker, see Fox (1987).

25. See Section 2.6 for an account of ontological dependence applicable to properties.

26. I am told that the criticism was originally due to Jack Smart.

27. The unknowability of categorical properties will later be developed as one of our principal objections to an ontology of passive matter, wherein all properties are ultimately categorical. The point was argued forcefully in Lierse (1996).

Part Three

Scientific Explanation

4

Realism and Essentialism in Science

4.1 ESSENTIALIST REALISM

Scientific essentialism provides a metaphysical foundation for scientific realism, and this metaphysic has been developed with precisely this aim in mind. The form of scientific realism that is underpinned by scientific essentialism may appropriately be called "essentialist realism." The thesis of essentialist realism will be developed in this chapter and the next.

The Basic Structural Hypothesis of scientific essentialism, according to which the world is objectively structured into hierarchies of natural kinds of objects, properties, and processes, is an objectivity postulate of just the kind that scientific realism requires. For, if it is true, then there must be objective ways in which things are classified by nature in each of these broad categories of existence. It implies that the world is naturally divided *vertically* into objects, properties, and processes and *horizontally* into kinds of objects, properties, and processes. Hence, if scientific essentialists are right, then it follows that the world itself has a rich store of objective facts – that is, of facts that exist independently of our knowledge, language, or understanding of the world. For at each point of intersection in this *natural kinds grid,* there is an objective fact. And all of these facts exist before we begin any of our investigations.

The question is, then, what kind of scientific realism is entailed by scientific essentialism? It will be argued here that what is entailed is a rich form of realism that any scientific realist should be happy to endorse. It implies realism about the kinds of things that nearly every scientific realist believes in. Yet it is a discriminate form of realism that, like the position known as "scientific entity realism," does not require belief in the theoretical entities of abstract model theories – that is, in such things as geo-

145

metrical points, perfect gases, possible worlds, perfect competition, sets, and numbers. Essentialist realism only requires realism about those theoretical entities that are postulated as being essentially involved in the causal processes described in our best causal-process theories. Some of these are things to which any scientific realist ought to be committed. Yet, essentialist realism is almost unique among forms of scientific realism in that it implies realism about the causal powers, capacities, and propensities of the things postulated as participants in these causal processes. It treats all such modal properties as genuine, not just as convenient fictions. Hence, essentialist realism is a position that no Humean scientific realist could possibly accept.

4.2 SCIENTIFIC REALISM

The standard argument for scientific realism (Smart, 1963) is presented as if it were just an argument from the best explanation. If the world behaves as if things like atoms and electrons exist, then the best explanation of this fact is that they really do exist. However, what is required is really an argument for the best kind of explanation. If one causal explanation is better than any other, then of course it is legitimate to prefer this explanation. Why else should we prefer to believe one causal story rather than another? But the alternatives to realistic causal explanations that have to be considered are not other explanations of the same kind, but explanations of completely different kinds. It is true that if the world behaves as if theoretical entities of certain kinds exist, then the best causal explanation of this fact is that entities of these kinds really exist. But what is needed to establish a case for scientific realism is not an argument for choosing between realistic explanations, but an argument that the best such explanations are better than any model-theoretic, or other, alternatives. Why should realistic causal explanations be preferred when there are other kinds of explanation available?

The main reason why the causal explanations are generally preferable is that the proposed alternatives to realistic causal explanations are usually just disembodied ghosts of such explanations. There were, for example, several different theories on offer a few years ago concerning the cause of the hole in the ozone layer. One of these, which traced the causal chain back to CFCs in the upper atmosphere, turned out to be the best explanation. Moreover, its superiority over the other proposed explanations became increasingly clear as further evidence was gathered. It was rational, therefore, to believe that the hole in the ozone layer was produced in the way described in this explanation, and increasingly irrational not to believe it. But belief in the explanation involves belief in the causal process, and therefore in the existence of the entities postulated as being involved in this process. However, there were no non-realist explanations of the cause of the hole in the ozone layer to compete with any of the proposed realist ones. A conventionalist construal of a realist explanation, such as "Without commitment to whether there are any ozone-destroying causal processes occurring in the upper atmosphere of the kind described in the preferred realist explanation of the ozone hole, there is certainly an ozone hole up there of the scale and kind that would be produced if these processes were occurring" is obviously phoney. It is not an alternative explanation that can sensibly be contrasted with the realistic one from which it was derived. It is an epistemically ultra-conservative opinion about the best explanation. Or, if it is not that, then we must ask whether anything would satisfy the conventionalist of the reality of the supposed cause of the hole in the ozone layer. If not, then the ultra-conservatism, and the doubt, are pathological, and there is no point in further debate. If realism about causes will not be accepted by a conventionalist, however good the causal explanations that might be provided are, then an argument from the best of all possible explanations will not be persuasive.

The same goes for idealism. There are no idealist explanations of the hole in the ozone layer with which realist explanations might be compared. Of course, an idealist will not be persuaded of the existence of the causal processes described in the best explanation. But then the idealist does not even believe in the existence of the hole in the ozone layer. So there is not much point in trying to construct an argument from the best explanation for use against an idealist. An argument for realism about theoretical entities is straightforwardly an argument from effects to causes. However, its scope certainly needs to be clarified. Smart, who developed the argument in the first place, thought that it applied to things such as atoms and molecules, and also to sets, but did not apply to properties or relations, unless properties could somehow be construed as sets.[1]

I have argued in a number of places (for example, in Ellis, 1987 and 1990) that the standard argument for realism also applies to properties, to forces, and, by implication, to causal powers. For these things all have essential roles in causal process explanations. Therefore we have to be realists about these entities too if we are to believe in these processes because any real effect obviously requires a real cause. On the other hand, I argued, the argument does not provide a satisfactory basis for realism about sets, or other theoretical entities of a non-physical nature, such as numbers, geometrical points, propositions, possible worlds, or such idealized objects as perfectly reversible heat engines. For these entities are not postulated as participants in causal processes, and scientists are not required to believe in them in order to accept the theories in which they occur. The relevant theories do not, and are not intended to, provide causal explanations.

Bigelow and Pargetter (1990) have defended a very strong form of scientific realism based on the standard argument for realism about theoretical entities. They call it "introverted realism." Like other scientific realists, introverted realists believe that the most successful theories in science are probably true, rather than, say, convenient fictions, or useful instruments for prediction. Consequently, they believe in the same kinds of things that most other scientific realists do. But they do so, as they say, reflectively (1990, 1998). That is, they not only believe that the theories that presuppose or assert the existence of these entities are probably true, they hold that if they are true, then they correspond to an independently existing reality. Only correspondence with such a reality, they hold, can explain their continuing success. However, it is not introverted realism that distinguishes their metaphysical position from that of most other scientific realists. For introverted realism, as they define it, is more or less standard scientific realism. There are very few scientific realists who are not also correspondence theorists about truth. What is most striking about their position, and what distinguishes it most clearly from other realist metaphysics of science, is their ontology. It is an ontology that includes many different kinds of things besides the usual physical entities and their first-order properties and relations. Most other scientific realists would not be willing to embrace them all.

To derive their ontology, Bigelow and Pargetter argue that the continuing success of a theory – any kind of theory, implies its (probable) truth, and that its truth implies the existence of an appropriate, independently existing, truth-maker. The truth-maker for a theory might consist of objects of various kinds, having various properties, and standing in various relations. If it does, then these properties and relations must all exist and be instantiated as the theory says. The truth-maker for a theory might, on the other hand, be some system of relations holding between other (lower degree) properties or relations, none of which is instantiated in the actual world. For such a theory, the truth-maker is a system of relations among universals and is therefore itself a complex universal. Bigelow and Pargetter's truth theory is therefore not just the correspondence theory. It has implications for ontology that no ordinary correspondence theory has.

Let us call Bigelow and Pargetter's argument that success implies truth, and that truth implies the existence of a truthmaker, "the strong argument for realism about theoretical entities." What makes Bigelow and Pargetter's ontology so different from, and so much richer than, the ontologies of most other scientific realists, is their uncompromising use of this strong argument for realism about theoretical entities. They use it right across the board – in logic, mathematics and semantics, as well as in science.

The ontology most commonly accepted by scientific realists is derived from an ordinary correspondence theory of truth. It combines realism about the postulated entities of our best physical theories with nominalism about their postulated properties and relations. This is the theory Smart (1963) once advanced, although he has now come around to accepting that there are determinable universals of various kinds.[2] In Ellis (1987, 1990), I defended a form of scientific realism that restricts the application of the standard argument for realism about theoretical entities to causal process theories,[3] and simultaneously extends it to embrace realism about the properties and relations postulated in such theories. Real things, I argued, must make a difference to physical causal processes.

Bigelow and Pargetter's ontology is not similarly restricted. On their theory, properties and relations can be real, even though they are not instantiated (or at least not instantiated in the actual world), and consequently they do not require real things to actually make a difference to what happens. If the theoretical entities we are considering are ones that feature in our best theories, whether these theories are physical, logical, semantical, or mathematical, then, by Bigelow and Pagetter's their criteria, there is a good case for believing in them. The theories from which we may derive our ontology need not be causal process theories.

With the exception of modal properties, such as causal powers, Bigelow and Pargetter generally refuse to accept any nominalist or conventionalist stratagems to avoid unwanted ontological commitments, as nearly all other scientific realists do. They are, as one of the authors once said to me, "the entities' friends." If, in our best scientific, logical, or mathematical theories, references are apparently made to entities of various kinds, then the best explanation for the success of these theories is usually just that they are true. Therefore, they argue, the referring terms in these theories must refer to real things, and the properties and relationships that are supposed to hold of, or between, them must be real. Hence, they conclude that most of the entities apparently referred to in our best scientific, logical, and mathematical theories are real. That is, most of the properties and relations ascribed to them must be genuine universals – except, of course, if the property happens to be dispositional.

149

By excluding dispositional properties from their basic ontology, Bigelow and Pargetter close off an important range of options for developing a theory of natural necessity. For if ontologically basic dispositional properties do not exist, and so cannot be the source of natural necessity, then there is nothing in the actual world that would ground this modality. Consequently, Bigelow and Pargetter are forced to go off-shore. The modality must depend on what exists in other possible worlds, and how the actual world is related to these other worlds.

While the energy and agility with which Bigelow and Pargetter pursue this outlandish hypothesis is admirable, I think it is fundamentally mistaken. If natural necessities are not grounded in the actual world, but in external relations between causally distinct worlds,[4] then natural necessities cannot necessitate. What does it matter how the actual world is related to other possible worlds, or as Bigelow and Pargetter probably should say, how the complex universal that the actual world instantiates is related to other complex universals that the actual world does not instantiate. How could such a relationship possibly necessitate anything? It seems to me that if all the possible worlds are in themselves non-modal, they will be so however they might be related to each other.

4.3 CRITIQUE OF THE STANDARD ARGUMENT FOR REALISM

The standard argument for realism about theoretical entities is clear, simple, and highly persuasive. It is most persuasive when applied to causal process theories – theories that purport to describe the processes that are supposed to cause the events to be explained. For if a causal process theory does successfully account for the kinds of events that are to be explained, there can be no doubt that the world behaves as if such processes are occurring. Moreover, if the postulated causal processes are supposed to involve things of certain natural physical kinds having certain essential properties, and if the causal process explanation of the events to be explained depends on there being things of just these kinds, then the world certainly behaves as if things of these kinds existed, and consequently have the properties that are essential to them. However, the classical argument for realism about theoretical entities is not always so persuasive. For causal process theories are not the only kinds of theories in science, and it is sometimes quite implausible to say that the world behaves as if the entities that are postulated in some of these other kinds of theories exist.

Consider, for example, the abstract model theories of classical thermodynamics. In the theory of the heat engine, the theoretically most important processes are adiabatic and isothermal expansions and compressions. But whether such processes actually occur in nature, or can be made to occur in the laboratory, is unimportant. The theoretically most important kind of heat engine is the Carnot cycle, for this is the heat engine that has maximum theoretical efficiency for the temperature limits between which it operates. Again, it does not matter whether such heat engines exist in nature, or can be manufactured. So it can hardly be said that the world behaves as if such heat engines existed. In fact, the world manifestly behaves as if there were no Carnot cycles. There is even good reason to believe that there cannot be any Carnot cycles in nature. Therefore the standard argument for realism about theoretical entities cannot be applied to argue that Carnot cycles exist. On the contrary, it can be argued that since the world behaves as if Carnot cycles do not exist, it is reasonable to suppose that this is the case.

Analogous reasoning would lead one to similar conclusions about the theoretical entities of other model theories. The standard argument for realism implies that there are no inertial systems, no perfect gases, no black body radiators, no geometrical points, no propositions, and so on. For these theoretical entities are all fairly obviously idealizations of ordinary things such as observational standpoints, gases, colored radiators, objects in space, and sentences. They are constructed in just the same kind of way as the processes of isothermal and adiabatic expansion are constructed as idealizations of ordinary processes of thermal expansion. Moreover, the world certainly behaves as if there were no things of any of these ideal types. Therefore, by the standard argument for realism about theoretical entities, we should conclude that they do not exist.

Bigelow and Pargetter's strong argument for realism about theoretical entities is much more widely applicable than the standard argument. It is meant to apply with equal force to cases just like the Carnot example. Indeed, it is meant to apply to the theoretical entities of all successful abstract model theories, including those of set theory, number theory, geometry, and Lewis's (1973) "posible worlds" semantics. It is not restricted in its application to causal process.theories, as I think it should be. Consequently, Bigelow and Pargetter must by their own argument be committed to the real existence of suitable truth-makers for Carnot's theory. For this theory has survived for 170 years or thereabouts,[5] and by any reasonable criteria of success is very successful. So, presumably, they must consider that there is a property of being a Carnot cycle operating between given temperature limits, and the processes of adiabatic and isothermal expansion and compression must occur in any engine that has this property, just as the theory prescribes.

151

Given what Bigelow and Pargetter have to say about things such as propositions, sets, numbers, geometrical points, possible worlds, and the like, which are the theoretical entities of other abstract model theories, it is not hard to anticipate what they will say about the property of being a Carnot engine. It is a universal, presumably a structural universal, that happens not to be instantiated in the actual world. According to their theory of universals, the fact that this universal is not instantiated does not matter. Universals can exist, even though they are not instantiated, provided they are instantiated elsewhere, in some other possible world.

This reply has much to recommend it. For it provides an answer to a question that would otherwise be very difficult to answer namely, why should some model theories be so much better than others? What metaphysical explanation can we give of their success? It is not – clearly not – that they offer better or more accurate or more comprehensive descriptions of the actual world. They do not. No real heat engine is even a close approximation to being a Carnot engine. The answer it offers is that good model theories work because they are dealing with real properties and relations – properties and relations that are genuine universals, not just human inventions. And these genuine universals are important ingredients in the structure of reality.

Still, there is something implausible about reifying the property of being a Carnot engine, and this is something that Bigelow and Pargetter may not wish to do. They may not wish to do so, not only because it is intrinsically implausible, but also because there cannot, even in principle, be a perfectly reversible heat engine. For, given the laws of entropy, it would have to operate infinitely slowly to be perfectly reversible, and therefore not really operate at all. A Carnot engine is an impossible object, and so cannot exist in any possible world. However, it is not clear what viable alternative Bigelow and Pargetter have in this case. It is not clear what other explanation they can offer of the continuing success of Carnot's theory. If they are not willing to insist that Carnot's theory is true, and hence must have a truth-maker, then they would be forced to admit that there are good theories that are good for reasons other than their truth. But this move is dangerous. It might be prove to be fatal to their whole program. If idealized things such as Carnot engines do not belong in the Platonic heaven Bigelow and Pargetter evidently believe in, why should other idealized entities, such as geometrical points, Euclidean figures, sets, and propositions, be admitted? Why treat the idealizations of mathematics and logic any differently from those of physical theory? Why should one believe in any of them?

Nevertheless, I think Bigelow and Pargetter have an important point. How else is one to explain the success of a very successful model theory? I think there is another answer that is metaphysically more satisfying than Bigelow and Pargetter's. Generally speaking, although probably not in all cases, model theories abstract from reality in order to focus on the essential nature of some kind of process or system of relations. What is essential to the nature of a heat engine for example, is the conversion of heat into useful work. In any ordinary heat engine, vast quantities of heat are lost by conduction, radiation, and convection. But it is not essential to the nature of a heat engine that such losses occur. What is essential is that the working substance be expanded at a higher temperature than that at which it is compressed, so that there is a net gain in the work done. To construct his model of an ideal heat engine, Carnot envisaged two ideal kinds of processes – isothermal and adiabatic processes. In the case of an isothermal expansion, all of the heat required to drive it is externally supplied, and ideally none is wasted. In an adiabatic expansion, all of the heat required to expand the gas is internally supplied, and again, ideally, none is wasted. Both kinds of processes are theoretically perfectly reversible. Therefore, any cyclic process that consists wholly of such processes would also be perfectly reversible. The Carnot cycle is such a process.

It is important to note that the point of idealizing in this case is not to simplify, but to eliminate what is not essential. The aim is to understand the essential nature of any continuing cyclic process of extracting work from heat. Carnot's theory was successful precisely because it did this. Similar considerations apply to other idealizations in science. Consider space-time theory. Ordinary events generally have a somewhat vague location in space and time. They do not occur in an instant, and they occur in a finite, and usually not very well-defined region of space. But vagueness is not essential to the concept of an event. Hence the time intervals and the regions occupied by events are of no importance from the point of view of a theory of space-time. It is the system of relations between events in which we are interested. So we envisage ideal events, which are neither spatially nor temporally extended, and proceed to construct a theory of the relations between such events. The theory is a good one to the extent it explains the actual relations that exist between real events.

I am not clear what the status of points is supposed to be on Bigelow and Pargetter's theory. But given their unrestricted use of the strong argument for realism, they should certainly be realists about them. Moreover, they should not seek to reduce them to any ontologically more plausible things such as numbers or sets of numbers, which, on one occasion, they seemed inclined to do (p. 364). A point is fairly clearly not a universal, although for every point in space or space-time there is plausibly a property of being located at that point. There is perhaps even a property of being a point. But individual points in space and space-time would appear to be particulars rather than universals. Indeed, they are suspiciously like bare particulars, which makes them ontologically very implausible. The most plausible view – that points are just theoretical objects, having no claim to be objectively real – is not available to Bigelow and Pargetter, because, to endorse this view, they would have to admit that our theories of space and space-time are theoretical models of reality rather than objectively true descriptions of it. But once they admit this possibility, their unfettered use of the strong argument for realism about theoretical entities must be brought into question.

Bigelow and Pargetter describe their position as *scientific Platonism*. The name is well chosen, because their theory is evidently a blend of scientific and Platonic realism. It is Platonic, because their theory of universals is Platonic. According to their theory, and contrary to Aristotelian theory, universals may exist uninstantiated, or, to be more precise, their existence does not depend on "whether or not they are instantiated by the fickle physical phenomena around us" (p. 365).

4.4 REALISM AND THE HUMEAN SUPERVENIENCE THESIS

Bigelow and Pargetter are realists about possible worlds. But the possible worlds they believe in are all categorical. They argue for the categoricity of worlds on the basis of Lewis's Humean Supervenience thesis – that all modal properties, including all dispositional properties, necessarily supervene on non-modal properties. As we shall see, there are many good reasons to reject this thesis. It is plausible only if one does not think too hard about what the ontologically most basic properties in nature might be. If many of these basic properties are dispositional, as I believe they are, and if these dispositional properties can be shown to be the truth-makers for the causal and statistical laws of nature, as I believe they can be, then by Bigelow and Pargetter's own argument for realism about theoretical entities, we ought to believe in them.

As a result of their uncompromising use of the strong argument for realism about theoretical entities, Bigelow and Pargetter are committed to the existence of many different kinds of things. They not only believe in things such as atoms and electrons, as all scientific realists do, but also in properties and relations (first-order universals); in properties of having, or

154

standing in, first-order properties or relations (second-order universals); in properties of, or relations between, first-order properties or relations (second-degree universals); in compound universals; in structural universals of various kinds, including grand universals, which are instantiated, if at all, by whole worlds; and in a complex range of higher level universals that are mainly the concern of mathematicians. Yet, strangely, they do not believe in dispositional properties, except as supervenient on categorical properties and laws of nature. As they put it, "we could not have any difference in modal [including dispositional] properties *unless* there were a difference also in nonmodal properties" (p. 175). For how a thing is, they suppose, tells us nothing whatever about how it is disposed to behave.

It follows from the Humean Supervenience Thesis that there are no causal powers, capacities, or propensities that exist as properties in their own right, since these are all modal properties. Any differences in the causal powers, capacities, or propensities of things would have to be grounded ultimately in some underlying structural or other non-modal differences. However, most of the most fundamental properties in nature would appear to be dispositional. The gravitational mass of a body is its intrinsic power to generate gravitational forces. The inertial mass of a body is its intrinsic capacity to resist acceleration by forces. The charge of a body is its intrinsic power to generate electromagnetic forces. The half-life of a fundamental particle is a measure of its intrinsic instability, which depends on its various intrinsic decay propensities.

Bigelow and Pargetter's ontology is thus a curious mixture of Platonism and Humeanism. There are all these universals that exist, and are the same in all possible worlds. But the world itself is a world of things whose only intrinsic and fundamental properties are categorical. If anything has a dispositional property, then this is a property it has by virtue of its categorical properties and the laws of nature, and the latter might well be different in other possible worlds. Hence, dispositional properties are not universals, and so cannot be admitted to primary ontological status. The fundamental particles must therefore be distinguished ontologically from each other, not by their causal powers, capacities, and propensities, as they currently are in practice, but ultimately by their primary Lockean sorts of qualities such as shape, size, structure, and other passive qualities. The various dispositional properties of the fundamental particles, must, according to Bigelow and Pargetter's theory, depend ultimately on the intrinsic, but at present unknown, constitutions of these particles.

I draw attention to this point because their commitment to a Humean ontology of actual existence occurs at an important stage in their argument, and without it their theory of laws would be untenable. They admit (p. 174) that the correspondence theory of truth is compatible with the existence of modal primitives. They reject this possibility, however, because they think that "modal properties must be supervenient on non-modal properties" (p. 174). Consequently, they restrict their attention in what follows to "theories which aim to characterize possible worlds without appeal to modal primitives" (p. 175).

This step is crucial because Bigelow and Pargetter require that there should be a Hume world corresponding to any given world. Their theory, and their logic, of natural necessity both depend on this assumption. The Hume world vis à vis any given world is a world exactly like the given world, containing the same things with the same first-order properties, and standing in the same first-order relations to each other, but without the causal connections or other relations of natural necessity that distinguish laws from accidental generalizations (p. 280). Every first-order thing that exists in our world is supposed to exist in the corresponding Hume world, and everything that happens in our world is supposed also to happen in this other world, but not by necessity. What happens in the Hume world happens purely by chance. However, if the standard dispositional properties of the fundamental particles and fields are their essential properties, as Fales (1986, 1990) argues, then it is logically impossible for such particles and fields to exist in a Hume world. But if there were no particles or fields like ours, then there could be no atomic or molecular structures like ours, and hence no DNA or other organic materials from which human or other living things could be constructed. Consequently, a Hume world would have to be a totally different kind of world. For it could not contain any of the kinds of things that exist in our world. If a Hume world that would look superficially like ours to any trans-world being (if there could be such a thing) could exist, then it would be intrinsically very different from the actual world. It would differ from the actual world all the way down to the most fundamental kinds of things that exist in these two worlds. This Hume world would be a globally counterfeit world containing none of the kinds of things that exist in this world.

One can take away the laws of nature, and leave everything in the actual world the same (except that everything that happens now happens only by chance), only if the laws of nature are not grounded in the actual world. If, however, they are grounded in the causal powers, capacities, and propensities of the fundamental physical kinds that exist in the actual world, as scientific essentialism asserts, then one cannot take away the laws of nature without also taking away the kinds of things they are concerned

156

with. Belief in the existence of a Hume world therefore depends on the rejection of essentialism. If an essentialist theory of causal and statistical laws is correct, and the dispositional properties of the fundamental physical kinds are the truth-makers for such laws, then it is logically impossible for there to be a Hume world containing the same kinds of things as those that exist in our world.

Perhaps the strangest thing about Bigelow and Pargetter's strong argument for realism about theoretical entities is that it should have led them to the very position they choose to ignore. They are anti-reductionist about most theoretical entities, and happily embrace such things as possible worlds, sets, and propositions. Yet they baulk at dispositional and other modal properties, and so are committed to reducing the intrinsic dispositional properties of the fundamental natural kinds to categorical properties and laws of nature. Why? Why not abandon the Humean Supervenience Thesis and build natural necessities into the world?

4.5 CAUSAL PROCESS REALISM

If Bigelow and Pargetter's introverted scientific realism is unsatisfactory, and Smart's naive scientific realism is too undiscriminating, then how can we improve upon these theories? What sorts of things should scientific realists believe in, and why? In earlier publications (Ellis, 1987, 1990), I sought to limit the scope of arguments for realism about theoretical entities to those involved in what I called "causal process theories." I argued that only those entities that are essentially involved in causal process explanations of physical phenomena should be admitted into one's ontology. Let us call this position "causal process realism." On this theory, if a physical causal mechanism is postulated to explain something, and it provides a good explanation – one that is better than any other currently available – then it is reasonable to believe in the reality of the proposed mechanism. Moreover, if the proposed mechanism assumes the existence of entities having various properties or structures, and these properties or structures are appealed to in the explanation that is given, we should believe that these properties or structures exist too. A causal process realist must therefore be a realist about certain kinds of properties and relations, as well as about the things that are supposed to possess them.

Specifically, it will be argued (in Section 6.9) that this requires realism about the causal powers of the theoretical entities that are postulated. Without them, or something else that would do the job, they could not be effective in the ways they are supposed to be. Hence, causal process realism turns out to be a species of ordinary causal realism. Causal process realism also implies realism about any categorical structures and properties of the things involved in the proposed mechanisms of causation, and in any structural properties or relations on which they might be ontologically dependent. But this, I suppose, is fairly uncontroversial.

In Chapters 6 and 7, we shall consider further what the argument for realism about causal powers and the like entails. It does, as we shall see, have some very important implications for ontology and also for the theory of laws of nature. But before we do so, there is one other implication of causal process realism that needs to be examined. Causal process realism not only implies realism about the participants in causal processes, about their structures, and about the *causal powers* that are exercised by them; it also requires realism about the various *causal processes* involved in an accepted causal story. These causal processes are no doubt displays of the causal powers, capacities, and propensities of the participants in the circumstances in which they happen to exist. But they are also parts of the causal process we are seeking to analyze, and so cannot be ignored.

In much of the literature of the philosophy of science, causal explanations are supposed to be nothing more than subsumptions of phenomena under causal laws. Something like this might perhaps be acceptable if the laws were genuinely causal and the phenomena were interpreted as the outcomes of genuine causal processes. But the standard view seems to be the Humean one that causal laws are just universal generalizations of some kind, which (it will be argued in Chapter 6) is unacceptable. Consequently, the standard theories of scientific explanation do not do the job one would expect of them. Thus, Hempel's covering law model, and its probabilistic variants, do not account for the interpretative nature of most scientific explanations. That is, the phenomena to be explained do not need to be interpreted as manifestations of causal processes, whether obvious or hidden. Instead, they are just to be subsumed under laws. We do not need to understand what is really going on behind the scenes. Explanation on these models has just the modest Duhemian aim of "summarizing and classifying logically" the experimental laws (Duhem 1954, 7). It does not, as Duhem says a genuine explanation should, aim at "strip[ping] reality of appearances covering it like a veil, in order to see the bare reality itself."

The statistical relevance models of explanation (Salmon, Jeffrey, and Greeno, 1971) are no better from this point of view. For they do not demand anything more than relevant statistical correlations. On these models, explanations are not required to tell us anything about the natures or structures of the entities involved in what is to be explained, or to describe what is really happening in the cases we are considering. It is enough if phenomena of the kind to be explained can be correlated statistically with some kind of prior occurrences, thus either increasing or decreasing the probability of the phenomena's occurring. No theory about the natures of the entities involved is required. It is as if these theories of explanation were specifically designed as accounts of explanation in fields of empirical research that are not concerned with the underlying causes of the events to be explained.

These theories of explanation have their roots in early positivism, and may appropriately be called "positivist theories of explanation." They are theories which are deeply embedded in the philosophical tradition that stems from Mach. In this tradition, science is seen as being concerned only with the description and prediction of nature, and if explanation has a role in science, then it is related to these aims, and only to these aims. Science is not seen as being concerned with the discovery of causal mechanisms or with the realities that may be supposed to lie behind the appearances. Nor is it seen as being concerned with discovering the essential natures of things – that is, the intrinsic properties and structures of things in virtue of which they are the kinds of things they are, and behave in the kinds of ways they do. According to the positivist tradition, the supposed essences of things are occult powers, and hence no empirical scientific explanation could ever refer to them. To the extent that a scientific theory seems to lead to discoveries about the essential natures of things, its claims are metaphysical and illegitimate – as Duhem argued in *The Aim and Structure of Physical Theory* and as Mach consistently maintained throughout his life.

The strong conclusions of Mach, Duhem, and other early positivists about how little science can tell us about what reality is like are clearly unacceptable to scientific realists. For scientific realists are precisely those who consider science to be the systematic investigation of the real world and its findings to be genuine discoveries about the nature of this reality. For them, science is the key to ontology. But scientific realists have not – or at least not until quite recently – systematically developed a theory of explanation of their own. R. Bhaskar (1978) did, and W.C. Salmon (1984) has now come to appreciate the incongruity of holding a realist theory of science together with a positivist theory of explanation. Mostly, scientific realists have just gone along with the Hempelian model, or something like it, without appreciating its inappropriateness. It is not that the Hempelian model is strictly incompatible with scientific realism. It is rather that it comes as something of a surprise to learn that a Hempelian is also a scientific realist. A theory of explanation that requires only laws and initial conditions as a basis for explanation, and does not demand of a scientific theory that it should reveal anything about the causes of phenomena, would not seem to require realism about any of the theoretical entities that might be postulated.

A theory of scientific explanation which does make such a demand – that is, a theory which demands that an explanation should tell us something about the underlying causes of phenomena – would require realism about those causes, because real phenomena could hardly be produced by fictitious ones (Ellis, 1987). Call such a theory "a realist theory of scientific explanation." Then this is the kind of theory of explanation in science you would expect most scientists, and all scientific realists, to hold. You would expect them to say that explanations of this sort not only exist, but are at least typical of, and perhaps even the only genuine, scientific explanations. Certainly it would follow that the theoretical terms referring to supposed hidden causes occurring in such explanations must be supposed to refer to real things, if these explanations are to be accepted. Not only are most scientific explanations concerned with the underlying causes of phenomena, and are thus realist in character, many are, I shall argue, also essentialist. That is, they are concerned with the essential natures of the kinds of causal processes they seek to explain.

Essentialist explanations of causal processes of various kinds are possible, of course, only if the processes to be explained are members of natural kinds. For only such processes can have essential natures – properties or structures that make them processes of the kinds they are. All causal processes are no doubt natural processes. But not every natural process belongs to a natural kind. Consequently, not all causal explanations are essentialist. Explanations of particular events in history, geology, or evolution, for example, are causal explanations, and their acceptance requires belief in the causal mechanisms proposed to explain them. But they are not essentialist explanations because they do not seek to lay bare the essential nature of anything. Facts about how things are intrinsically disposed to behave are just taken for granted, and the problem is seen as being to explain how things with these dispositional properties happened to combine to produce the effects to be explained.

What kinds of processes, then, belong to natural kinds? Processes that depend on the accidental facts of history manifestly do not. I believe that it is one of the primary aims of physical science to isolate and describe the natural kinds of processes that occur in the world, and to discover their essential natures. There is no better illustration than this than in the theory and practice of chemistry.

Every chemical theory about every kind of chemical reaction is an attempt to describe the essential nature of a natural kind of process. Consider, for example, the process by which cupric sulfate (or bluestone) is made commercially. This process involves metallic copper (which is essentially just Cu), sulfuric acid (which is essentially H_2SO_4), and oxygen from the air (which is essentially O_2). In practice, metallic copper is sprayed with hot dilute sulfuric acid in a chamber through which a current of air is flowing. The reaction that normally occurs under these conditions is essentially:

$$2Cu + 2H_2SO_4 + O_2 = 2CuSO_4 + 2H_2O$$

Of course, other reactions might also occur under these conditions – for example, if the copper or the acid is impure, in which case there will be several different reactions occurring at the same time, or if there is insufficient oxygen, in which case no reaction, or a different reaction altogether, might occur. But this equation, we may reasonably suppose, describes the essential nature of the reaction that is of primary commercial interest. There is no other kind of reaction that could possibly have this structure. Nor could any given reaction lack this structure, yet still be a reaction of this kind. There are of course other reactions that result in the production of cupric sulfate – for example, ones involving cuprous or cupric oxide. But these are different reactions, and the chemical equations that describe them are different.

All such theories are essentialist. If, for example, you ask what makes the cupric sulfate chemical reaction one of the kind it is, you will learn that it is the fact of its being a reaction involving just these kinds of molecules in just this kind of way. For this is the fact that explains its similarities and differences from all other reactions. It explains the combining weights of all of the reactants and products, the conditions under which they react, and (with auxiliary information) the readiness of the reaction to proceed in the direction in which it does. If you ask what distinguishes this chemical reaction from any other, you will quickly learn that there is no other reaction that has this structure. Having this structure is both a necessary and sufficient condition for anything's being a reaction of this kind. Moreover, this is true, not because of any arbitrary linguistic conventions, or ways in which the events are perceived or evaluated by us, but because this kind of reaction is categorically distinct in nature from any other kind of reaction. Its differences from all other kinds of reactions would exist whether or not human beings ever existed.

If the chemical theory describing this reaction is true, then it is necessarily true of all and only reactions of this kind. Often, kinds of chemical reactions do not have proper names. This is a fact about language and ultimately about us. Kinds of material objects are salient to us in a way in which kinds of processes are not. But there is no ontological reason why kinds of processes should not be named. Let us coin the general name "bluestone-gen" for the chemical process described for the commercial manufacture of cupric sulfate. Then all process of the kind bluestone-gen must have the structure described in the chemical equation, not because this is what it means to be a process of the kind bluestone-gen, but because this is what it is to be such a process. Nothing could be a chemical reaction of this kind, if it were not a process of the sort described by the equation. It does not matter what names, if any, we might have for chemical reactions of this kind. For the thesis has nothing to do with what is true in virtue of the meanings of words. It has to do rather with what is necessary in reality – that is, it is a question of what is really necessary.

The same is the case for all kinds of chemical reactions. Their identities as kinds depend on their chemical analyses. If litmus paper turns red as a result of being dipped in pH-neutral red ink, then however like in appearance, to the chemical reaction that causes litmus paper to turn red in acid solutions it is not an instance of this kind of reaction if the red ink is pH-neutral. To be an instance of this kind of reaction, it must have the same chemical cause and be brought about by a chemical process of the same generic kind.

When we think about natural kinds, we mostly think about natural kinds of objects or substances (animals of a given species or atoms of a given element). But science abounds with descriptions of natural kinds of processes. The kinds of chemical reactions are good examples of such processes, but there are many others from the physical sciences. The processes of electromagnetic radiation, pair production, radioactive decay, refraction, reflection, osmosis, diffraction, crystallization, fusion, and vaporization, to take some examples at random, are all natural kinds of processes. And it is now possible to say, or at least present a respectable theory about, what makes any given process a process of one of these kinds and what distinguishes processes of any given kind from those of any other. Even biology seems to have its share of natural kinds of processes. Meiosis and meitosis come to mind immediately. It is true that many of the natural kinds of processes whose natures we are now able to describe have never been named. Only the most theoretically important kinds of processes get named. But this is more a fact about language than the world.

It may be objected that the processes that I say are members of natural kinds are really just the ways in which things of different kinds are bound to behave, given their circumstances and the laws of nature. The kinds of objects that the chemical processes involve are members of natural kinds. Let that be granted. But why make the additional claim that the distinctive ways in which they may interact are processes belonging to natural kinds? It may well be granted that these interactions satisfy all of the criteria for membership of natural kinds. But isn't this sufficiently explained by the fact that the participants in these interactions are essentially distinct? In other words: Might not the distinct kinds of processes just supervene on the distinct kinds of substances and the laws of nature that govern them?

This would be a reasonable position to take, if (a) it could be shown that the kind-identities of substances were independent of how they are intrinsically disposed to behave, and (b) it could be explained satisfactorily why the laws of nature should discriminate between intrinsically different substances as thoroughly as they do. The first of these two claims, I argue, is false. It is the thesis of categorical realism, which was examined and rejected in Chapter 3. The second of these two claims is correct in its assertion that the laws of nature are sufficiently fine-grained to enable us to discriminate between chemically different substances (how else could we know them to be different?). But this would be most readily explained if the laws of nature were grounded, not in universal commands, but in the natures of the substances themselves (which is what a scientific essentialist says).

There is a close connection between natural kinds of properties and natural kinds of processes. For what makes a property the kind of property it is depends on how things having this property are intrinsically disposed to behave. A thing having unit mass, for example, must be disposed to behave as an object with unit mass. It cannot, qua object of unit mass, be disposed to behave in any other way. Of course, as well as having unit mass, an object may have, say, unit charge. In that case, it must also be disposed to behave as an object with unit charge. Therefore the behavioral patterns actually exhibited by objects of unit mass may be quite complex, depending on what other properties they have and what their circumstances are. And, to be fully understood, these behavioral patterns may need to be analyzed into the separate components that are due to their several properties.

But this is the way with processes – they can occur to the same things at the same time. An object that is being translated from one place to another may also be rotating, radiating heat and transmitting radio waves. Similarly, an object that has a certain property, and is thus disposed to behave in a certain way, may manifest some quite complex behavior that cannot be explained just by its having the property in question in the circumstances in which it exists. Its complex behavior may be due to its having a number of other properties whose several manifestations contribute to the complex behavior that is actually displayed.

The ways in which properties and processes compound in the actual world make the attempt to discover what kinds of properties things may have, and in what kinds of processes they may be involved, a very complex business. Science proceeds to unravel the manifest complexity of what occurs in the actual world by making and testing various hypotheses about what natural kinds of objects, properties, and processes exist, and what their essential natures are. In chemistry, this has been achieved by an investigative process that has involved the parallel development of a hierarchical chemical classification system (of elements and compounds), a theory of chemical identity adequate for this system (the atomic-molecular theory), an extensive knowledge of the chemical properties of the various substances, so classified, and sophisticated theories to explain the sources of these properties, and the kinds of chemical processes for which they are responsible.

The historical process by which all of this chemical knowledge and understanding was achieved was very complex, and involved many different kinds of investigations. It required the isolation and purification of the different kinds of substances and the systematic study of their manifest and dispositional properties. It needed the development of theories of chemical processes, adequate to explain the various kinds of chemical changes and reactions that can be seen to occur, compatibly with the underlying theory of chemical identity. And it required the discovery and explanation of the various kinds of chemical properties the different substances can have.

The kinds of investigations involved in this complex epistemic process are all strongly interdependent. Hypotheses about what kinds of substances exist may depend on theories of chemical properties or processes, and conversely. The point is well illustrated in the writings of Antoine Lavoisier. Lavoisier believed acids to be essentially oxygenated bases or radicals. Thus we have carbonic acid (CO_2), sulfurous acid (SO_2), nitrous acid (NO), nitric acid (NO_2), and so on. He argued, therefore, that what we now know as hydrochloric acid (HCl) – spirit of salt – must also be an oxide (X_mO_n). He called the postulated base X of this oxide "muriatic base," and the acid "muriatic acid," but lamented that "this acidifiable base adheres so very intimately with oxygen, that no method has hitherto been devised for separating them" (Lavoisier, 1789, 70).

But, complex as the scientific reasoning involved in these early investigations may be, it is clear that it presupposed the existence of natural kinds of substances, properties, and processes, and involved the development of theories concerning the essential natures of all of these natural kinds. Therefore, in the field of chemistry at least, this kind of reasoning must be regarded as fundamental. The process of acquiring chemical knowledge cannot possibly be understood if we do not know the nature of such reasoning.

4.7 ESSENTIALISM IN PHYSICS

To give essentialist explanations of physical processes, we must first seek to identify the natural kinds of things involved in them and to discover their essential properties. Consider, for example, the sort of explanation we might offer of the diffraction patterns that are observed when X-rays are bounced off a metal surface. First, it is assumed that the process by which this occurs is one that belongs to a natural physical kind. And because it has certain analogies to optical diffraction, the process is called "X-ray diffraction." The question, then, is what is the nature of processes of this kind. To answer this question, it is evident that we must proceed from theories about the nature of X-rays and about the kinds of surfaces from which X-ray diffraction patterns arise. For independent reasons, X-rays are identified as electromagnetic waves of very high frequency. Making this identification is already a very big step toward the construction of the required theory. For having made this identification, we may immediately conclude that X-rays must be wave-like in their interactions with each other, and particle-like in their interactions with other things. We may also conclude that X-rays may be polarized, and would be influenced in various ways by electric and magnetic fields. For it is essential to the nature of electromagnetic waves that they should behave and be influenced in these ways.

Second, the kind of surface that gives rise to X-ray diffraction patterns must be identified. For to explain the phenomena, we need to know what kinds of things the X-rays are interacting with. The second step, then, is to identify the nature of the surfaces that produce these patterns. The diffracting surfaces are assumed to be the plane surfaces of crystals. Once this second identification has been made, we are in a position to say a great deal more about what the process of X-ray diffraction must involve. We know from crystallography that a crystal must have one or other of a certain range of structures, depending on what kind of crystal it is. We know from quantum theory that the scattering must occur by a process of absorption and reemission of X-ray photons by the atoms located in or below the crystal surface. And we know from wave theory how the reemitted waves must interact with each other to produce interference and reinforcement patterns. It remains only to construct theoretical models of the various crystal structures and the incident and refracted X-rays and to demonstrate that such models are adequate to explain the various diffraction patterns that are actually observed, given the wavelengths of the X-rays concerned and the inter-atomic distances in the crystal lattices from which the diffraction patterns are seen to emerge.

To demonstrate the adequacy of such models, it is normally sufficient to show that the patterns to be explained are more or less what is expected, given that the process of X-ray diffraction is essentially as described. It is not necessary to be able to predict the phenomena precisely. For the real-life situation will be enormously complex. The incident X-rays will not be monochromatic or perfectly collimated, as the model is likely to assume. The crystal structure will almost certainly have flaws and contain impurities of various kinds, which the model will quite properly ignore. The model will ignore these details, because the aim of the exercise is not to save the phenomena exactly, but to describe the essential nature of the processes that give rise to the phenomena.

Theories concerning X-ray diffraction are likely to be fairly realistic. For the contingencies under which the phenomena of X-ray diffraction occur are likely to have a relatively small impact on the phenomena to be explained. Such is not the case for other natural processes. Consider, for example, the formal theory of the natural processes by which substances are absorbed in plant and animal nutrition. One of these processes is osmosis. The formal theory of osmosis makes use of the concepts of an ideal solution (which is essentially similar to the concept of an ideal gas[6]) and of a semi-permeable membrane (a membrane that is permeable to molecules of the solvent, but not to those of the solute). Now, many naturally occurring membranes approximate fairly closely to the ideal of semi-permeability for at least some of the kinds of solutions involved in these processes. But there are degrees of permeability and hence of semi-permeability, and these degrees can be modified by a host of accidental factors, including temperature, thickness, surface condition, presence or absence of grease, and so on. Therefore, perfect agreement of theory with experiment is simply not to be expected in this area, even if the solutions we had to deal with were always ideal ones.

Moreover, no real solution is ever ideal, and many are far from ideal. The forces of attraction between the molecules of the solvent, between the molecules of the solute, and between the molecules of solvent and solute are not all the same, as they should be in any ideal solution. And there are many other reasons why real solutions are not ideal. In fact, no solutions are even very good approximations to ideal solutions, except at low concentrations of solute. In this respect, they are very like ideal gases. Of course, there are more sophisticated models of solutions, which take into account more of the incidental factors that influence osmotic pressures – factors that depend on the specific properties of the solvents and solutes concerned. But, even so, no model yields strictly accurate predictions concerning osmotic pressures. However, this does not mean that the phenomenon of osmosis is not well understood. On the contrary, the essential nature of processes of this kind is certainly known. What is not known, or not known precisely, is what other factors, whether accidental or incidental, may influence it, or the extent of their influence.

The idealizations involved in the theory of osmosis do not differ in principle from the idealizations we have to make in constructing a model to explain the phenomena of X-ray diffraction. In both cases, we have to abstract from the accidental features of the systems we are dealing with – features that may influence what happens in them in unpredictable ways, or are not the focus of our concern. The solutions are not ideal solutions, and their behavior depends on the specific properties of the substances they contain and on the concentrations in which they are found. Moreover, the membranes through which osmosis occurs are never perfectly semi-permeable. But, as idealizations, they do not differ in principle from the perfectly collimated, monochromatic X-rays impinging on perfect crystals postulated to explain the phenomena of X-ray diffraction.

4.8 ESSENTIALISM IN BIOLOGY

Explanations having many of the characteristics of the essentialist explanations of physics and chemistry are to be found in most sciences. For all sciences are concerned with explaining kinds of behavior. Even if the kinds in question are not strictly speaking natural kinds, the objects involved are often sufficiently similar in their constitutions and exhibit sufficiently similar patterns of behavior for explanations of the sorts characteristic of the physical sciences to be discoverable.

Of course, the assumptions from which such explanations proceed will not, even if true, be necessarily true, if the kinds of events or processes to be explained are not natural kinds. Nevertheless, the postulates we make about the intrinsic natures of the things concerned in these processes might well, as a matter of fact, be true of the things that actually exist in the world. If so, then we can readily explain why things of the kinds in question must behave as they do. They must behave as they do because they are in fact constituted as they are, and all things so constituted must behave in this sort of way. But if the kinds we are dealing with are not natural kinds, we cannot argue that things of these kinds must be constituted as we suppose, or else they would not be things of the kinds they are. That is, we cannot argue that things of this kind must behave in this sort of way in every possible world.

Many biological explanations are straightforwardly essentialist, and most of the processes described by biochemists and microbiologists are processes belonging to natural kinds. The identities of these processes depend on the kinds of substances or structures involved in them and on the chemical interactions that drive them. The cells and cell structures are natural kinds of things, and their processes of growth and reproduction are ordinary causal processes. Like causal processes in any area, they can be thwarted or interrupted, and appropriate conditions of heat, nutrition, environment, and so on have to be met if they are to proceed. But the same is true of chemical processes generally, and I see no reason to think of microbiological or biochemical processes as being fundamentally different from the processes occurring in non-living matter. However, it is not clear how much of biology is concerned with the study of natural kinds. Aristotle believed that plants and animals are objectively classifiable into species. To explain these objective differences of kind, he supposed that the members of different species had their own characteristic sets of fundamental, species-determining properties – properties that no member of the species could fail to have, and that together comprise a set of properties that nothing else could have without being a member of the species. These were the species-determining or essential properties of the species.

Locke (*Essay*, Book III) attacked Aristotle's theory of natural kinds, arguing that the boundaries between species are not naturally sharp. According to M.R.Ayers (1981), Locke's criticisms of Aristotle's theory were motivated by the belief that

all differences are differences of degree, and everything is in principle indefinitely mutable. For all differences and changes are ultimately just differences and changes in the spatial quantity and ordering and motion of the parts of things (p. 255).

However, not everyone would agree with Ayers on this point. H. Kornblith (1993, Ch. 2) argues that Locke vacillates between three different points of view. There is the *strong nominalist* view, noted by Ayers; the more widely attributed *transcendental realist* view that natural kinds exist, but they, and their real essences, are unknowable; and the *empirical realist* (and also scientific essentialist) view that natural kinds exist, can be identified, and their real essences discoverable by scientific investigation. Whatever Locke's views may have been, there are certainly many differences between basic kinds of things that are not just differences of degree. This is one of the lessons of quantum theory. The differences between successive elements in the Periodic Table, for example, are fundamental ones that cannot be bridged by any intermediate cases. Hence we know that objectively distinct kinds of things really do exist. This much, at least, should be beyond dispute, although J. Dupré (1993) seems to think otherwise. What remains in dispute is what things actually do belong to natural kinds. What in particular is the status of biological species?

According to modern biology, the natural basis for a taxonomy, and hence the obvious place to look for essential properties, is in the genetic constitutions of plants and animals. But within a species, there is often a lot of genetic variation, and sometimes there are no sharp genetic distinctions between different species. Moreover, if one wishes to have a classification system that applies to extinct as well as to living species, as one surely does, and which can accommodate any new species that might be created by genetic engineering, as one reasonably might, then the problems of classification in terms of genetic makeup become even greater. If evolution occurs in the gradual kind of way that Darwin supposed, or if small changes in genetic constitution can be brought about artificially, then the distinctions between adjacent species – living, dead, or yet to be created – must ultimately be arbitrary.

For these and other reasons, many biologists and philosophers of biology say that they do not believe in natural kinds (J. Dupré 1993), or at least would deny that species are natural kinds (R. de Sousa 1984). Some argue that species are individuals, rather than kinds (M. Ghiselin 1974, 1987; D.

Hull 1976, 1978, 1981). Others (for example, M. Ruse, 1987,[7] T.E. Wilkerson 1986, 1988, 1993, 1995) allow that species are natural kinds, but not in the strict Aristotelian sense. Wilkerson argues that biological species are clusters of intrinsically similar natural kinds, and that members of biological species are consequently sufficiently like each other intrinsically for them to be amenable to most sorts of natural kind reasoning. Hardly anyone denies that the genetic constitution of a plant or animal growing in a normal environment determines many of its more obvious characteristics. They deny only that a satisfactory classification system can be based on this fact. From the point of view of genetics, the only strict biological natural kinds that have more than one member are kinds consisting of genetically identical twins, or clones. But this does not make for a useful taxonomy of plants and animals. I accept T.E. Wilkerson's view that biological species are more or less salient clusters of intrinsically similar natural kinds[8] – sufficiently similar in fact for it to be reasonable in most cases to think of them as, and reason about them as if they were, strict Aristotelian natural kinds.

Because of the messiness of biological kinds, and in order to develop a theory of natural kinds adequate for the purposes of ontology, I have broken with the tradition of using biological examples, and taken the various kinds of fundamental particles, fields, atoms, and molecules as paradigms. It does not matter if the resulting theory of natural kinds does not yield a satisfactory biological taxonomy. For my aim was to develop a theory of natural kinds and their essences applicable to what I take to be ontologically much more basic constituents of the world than biological species. If my doing so throws some light on problems of biological classification, then that is a bonus. Perhaps it does. Darwinian explanations of evolution of species are not essentialist. The same species could in principle have evolved in many different ways. It might in fact never be the case that different groups of animals of the same species evolved by different routes. But if such a discovery were to be made, the existence of coevolution would be a curiosity, rather than grounds for a distinction between species.

4.9 ESSENTIALISM IN PSYCHOLOGY

There are few, if any, explanations in psychology that are essentialist in character. But there is no good reason why such explanations should not exist. For although human beings are not strictly speaking natural-kinds, they are sufficiently like each other neurologically for natural kind reasoning about mental processes to be appropriate. U.T. Place and J.J.C.

Smart speculated in the 1950s about what these processes might be, although they were certainly not essentialists. They argued that such things as being in pain or having a sense impression of redness might really be brain processes. It is not logically necessary, they insisted, that they should be such processes. The identity in question, they said, is a contingent identity. Thus, they speculated that being in pain is, as a matter of fact, something like having "c-fibers" firing in one's brain.

This thesis became known as "the contingent identity thesis." However, the concept of contingent identity was never clearly articulated. It was said to be like the identity between the Morning and the Evening Star, between water and H$_2$O, and between temperature and average molecular kinetic energy. But there were no satisfactory arguments to show that such identities are contingent. Certainly, they are *a posteriori*. But as Kripke (1971) argued later, a-posteriority and contingency are not the same thing. If these identities hold in any world, then they hold in every world. Therefore they are not contingent identities, but necessary ones (as indeed all true identities are). Of course, these identities had to be discovered empirically. But this is a different matter. It does not show them to be contingent.

As we understand them, Place and Smart were speculating about the identities of mental processes. Sensations, they were saying, really are brain processes of some sorts. They might well have been right about this. But if they were right, then it is an essentialist thesis that they were right about. It is indeed just like the thesis that water is H$_2$O or that temperature is average molecular kinetic energy. It is not a contingent identity thesis that they got right, however, but an *a posteriori* identity thesis. Of course, in allowing for this possibility, I am not asserting that there are any kinds of mental processes that can be identified one to one with kinds of brain processes. For the identities of the mental processes of which we are subjectively aware might well depend on their gestalt features or on their functional roles. There might, for example, be many intrinsically different sorts of brain processes that a person would classify as painful – for example, because of the discomfort they occasion, even though these processes have no common features with which an individual's painful sensations could be identified. Moreover, the same mental processes might be the result of intrinsically different brain processes in different individuals. Indeed, it seems to me that this is likely, and hence that psychological essentialism is false.

George Bealer (1994) has an *a priori* argument for the falsity of psychological essentialism. It depends on his view that our modal intuitions are basically to be trusted. But, in trusting our intuitions, we are some-times led into a kind of stalemate – a situation in which our intuitions are in conflict. On the one hand, we wish to assert that there is an identity – for example, between water and H_2O, and so maintain that water is *necessarily* H_2O. On the other hand, we are strongly inclined to say that it might have turned out to be otherwise, and that in another world, it might ac-tually be otherwise. Hence, the alleged identity is only a contingent cor-relation. To resolve this conflict, and to swing the argument in favour of scientific essentialism, Bealer adopts Kripke's rephrasal strategy to show how both intuitions can be accommodated. Water is necessarily H_2O, but it is contingent on the clear liquid's we call "water," drink to quench thirst, use on the garden, and so on being water. Could be something else – for example, XYZ, and perhaps on Twin Earth it is XYZ. Bealer argues, how-ever, that this strategy fails for the supposed identity of being in pain and having c-fibers firing, even if the correlation turns out to be both unique and universal.

I think that Bealer is right to say this, and that he has demonstrated that there is a clear difference between the cases. However, I do not think it shows what he thinks it does. He thinks, wrongly in my view, that it shows that it is possible for something to be in pain, but not have firing c-fibers. It does show something like this. But what it shows has no bearing on the contingent identity thesis that was once so popular among Australian philosophers.

Let us agree, to get it out of the way, that it is epistemically possible that something could be in pain, but not have firing c-fibers. For all we know, it might be something else that is happening in our brains. Likewise, for all Socrates knew, water might have been something other than H_2O. However, we now know that it is not really possible. So, what is epis-temically possible is not an infallible guide to what is really possible. In fact, it is a very unreliable guide, as the history of science amply demon-strates. So the question is whether it is really possible for something to be in pain, but not to have firing c-fibers, or conversely. Let us also agree, this time for the sake of simplicity, that everything that is in pain in this world has firing c-fibers, and that nothing has firing c-fibers that is not also in pain. This is not a modal thesis, and it may well be true, whatever view we may take on the modal thesis that this holds necessarily.

I take the view that the world itself is one of a kind (see Section 7.9). It is a world of a natural kind characterized by a certain fundamental on-tology and by certain global properties and structures. In worlds of our kind, I would suppose that there are certain constraints on the kinds of beings that can exist. Not any kind of being that a science fiction writer cares to imagine is a being of a kind that could exist in our kind of world. Our imaginations can easily outrun what is really possible. For what is really possible is just what is possible in our kind of world. So the question is whether there could be a being in our kind of world that is in pain, but does not have firing c-fibers, or conversely. If not, then we may conclude that this is what being in pain really is in our kind of world. The identity thesis holds.

What Bealer has shown is that even if this were so, we might still be forced to agree that in another possible world a being might be in pain but not have firing c-fibers. For the two properties cannot be identified conceptually. Good, then let us call this sense of possibility "conceptual possibility," although we already have another name for it − "epistemic possibility." Then what Bealer has shown is that it is conceptually, or epistemically, possible for something to be in pain but not have firing c-fibers. But the mere epistemic possibility that something might be in pain but not have firing c-fibers does not demonstrate that it is really possible. That depends on what kinds of things and processes are capable of occurring in worlds of the same natural kind as ours.

4.10 CONSTRUCTING ESSENTIALIST EXPLANATIONS

All natural kind reasoning seems to depend on hypotheses about what natural kinds of objects, substances, properties, and processes exist, and theories about what their essential natures are. If we believe that K is a natural kind of substance, for example, and that x is an instance of this kind, then we will suppose that the intrinsic causal powers, capacities, and propensities of x, − and, if it is a complex substance, then its intrinsic structure as well − are properties it has necessarily in common with all other instances of this kind. For these intrinsic properties and this structure will immediately be identified as the essential properties and structure of the kind. Therefore, once the kind is known, the empirical task is to determine what the intrinsic dispositional properties and the intrinsic structure of any particular instance of the kind are.

Reasoning about natural kinds of properties and processes is similar. It proceeds from assumptions about the kinds of properties and processes that exist and what kinds of properties or processes are instantiated in the particular case, or cases, being investigated.[9] There are, however, some important differences between natural kinds of substances, on the one hand, and natural kinds of properties and processes, on the other. First, natural

173

kinds of properties and processes are typically spectral. That is, their instances are not all identical, qua properties or processes, but differ from each other quantitatively in one or more ways. Consequently, the essences of these kinds must be described quantitatively – as the measures of their influences in causal processes, in the case of causal powers and capacities, and as quantitative relations between causes and effects, in the case of causal processes.

Second, properties and processes are often hidden, or their effects swamped by other properties and processes. Indeed, some properties and processes rarely, if ever, occur so that they can be observed without interference, and often their effects cannot be measured directly. Consequently, to describe such processes, or the effects of such properties, it is often necessary to abstract from anything that can actually be observed to consider what would happen in the imagined absence of other factors that exist in the actual situation. As a result, descriptions of the essences of causal properties and processes are often abstract, and expressed either categorically, as statements about the behavior of idealized objects in ideal circumstances, or subjunctively, as conditionals about how real objects would behave, if they, and the circumstances of their existence, were ideal.

But although there are these differences between natural kind reasoning about kinds of properties and processes, on the one hand, and substances, on the other, the basic structure of the reasoning is the same.[10] Any instance of a natural kind of property or process is in principle as good as any other as a basis for generalizing to all members of the kind.[11] Indeed, as we shall see, it serves as a basis for the claim that all members of the kind must have certain effects (depending on the magnitude of the cause) or must proceed in such and such a way and have such and such effects (discounting the influence of other causes). That is, natural kind reasoning leads not only to universal generalizations about substances, properties, and processes, it implies that these universal generalizations are in some sense necessary.

The statements that describe the essences of the natural kinds of causal processes are a species of laws of nature. They are the causal laws. In Sections 6.9 and 7.8 it will be argued that these laws are necessary in the sense in which any statement ascribing an essential property to a natural kind is necessary. That is, they are really or metaphysically necessary. The theory of natural kind reasoning thus explains not only the process by which we may arrive at a knowledge of causal laws. It also explains the modal status of these laws.

174

NOTES

1. See Smart (1963, 1987) and Ellis (1987, 1990).
2. See his paper "Laws and Cosmology" in Sankey (ed. 1999).
3. As did Cartwright in her (1983).
4. I speak here about other possible worlds in the way that Lewis would, even though Bigelow and Pargetter specifically reject Lewis's concrete modal realism. My excuse is that Bigelow and Pargetter do the same.
5. Sadi Carnot's calculation of the maximum theoretical efficiency of a heat engine dates from 1824.
6. An ideal solution is often defined as any solution that obeys Raoult's law. Raoult's law states that the relative lowering of the vapor pressure of the solvent due to the solute is proportional to the mole fraction of the solute, and is independent of the temperature of the solution. However, this way of defining an ideal solution obscures its theoretical basis.
7. Ruse (1987) argues that biological kinds (species) are not natural kinds in the strict Aristotelian sense. Species do not have essential natures as Aristotelian natural kinds must have. On the other hand, biological species are not just nominal kinds either. We do not divide the animal or plant kingdoms into species just on the basis of similarities and differences that we happen to find salient, or convenient, or whatever. And where there are cases that look like borderline cases, we do not determine species membership just by fiat. Species are, Ruse argues, natural kinds of a looser sort. They are natural kinds because the distinctions between species depend on the existence of convergent criteria. Animals that belong together as members of the same species, according to one classificatory criterion, are typically found to belong together on the basis of other such criteria. Ruse (p. 238) supports Whewell's contention that "The Maxim by which all systems professing to be natural must be tested is this: that the *arrangement obtained from one set of characters coincides with the arrangement obtained from another set*" (Whewell, 1840, Book I, 521, his italics).
8. This is probably the most widely accepted view. Caplan (1981) is basically right to think of species as groups of animals or plants distinguished from other groups of animals and plants by the genotypic families to which they belong. However, there are no sharp distinctions between genotypic families either, at least not of the sort required for groupings into natural kinds. The correct view, I suppose, is Wilkerson's (1995) that for natural kinds we should require genotypic identity, not genotypic similarity. Hence, from the point of view of ontology, each individual must be supposed to define its own species, unless it just happens to have an identical twin or clone.
9. Several philosophers have argued that natural kind reasoning concerning properties and processes involves consiliences of inductions, allowing what appear to be different kinds of processes to be seen as processes of the same kind (Butts, 1977; Forster, 1988; Harper, 1989). I would agree that such consiliences of inductions do occur, and often signal important advances in conceptualization. However, I do not think they describe the kind of reasoning that is characteristic of subjects such as chemistry. I think Macnamara (1991) is right about that.

10. It is accurately described in Macnamara (1991) insofar as the reasoning is concerned with natural kinds of substances. I think exactly the same kind of reasoning is apt for natural kinds of properties and processes.

11. Macnamara's (1991) logic of induction depends crucially on his concept of an *anchoring common noun*. Specifically, he suggests that an inductive inference will be sound iff it is made with respect to an inductive predicate "P," which can be shown to hold of all members of an arbitrary sample S of things that are members of a kind defined by a given anchoring common noun 'C(M)'. The soundly drawn conclusion will then be "All C(M) are P." However, the concept of an anchoring common noun is not analytically defined. As far as I can see, it is defined with reference to the inductive inferences we commonly accept as sound – that is, with respect to our established inductive practices. A common noun is an anchoring common noun, he says, iff either: (a) It is the universally quantified common noun in the conclusion of an inductive inference delimiting the extension to which an inductive predicate is inferred to apply, or (b) it refers to the kind that by inductive inference is shown to have a kind C'(M') as a proper subset (p. 37). So the question of which nouns are anchoring common nouns depends on which inferences are inductive inferences and which predicates are inductive predicates, and these questions are themselves to be resolved with reference to our practices. So, presumably, "grue" is not an inductive predicate because we should not infer grueness to be true of the entire population of emeralds on the ground that it has been shown to be true of a proper subset of emeralds (cf. Macnamara's Definition 3, on p. 37). On the other hand, "emeralds" is an anchoring common noun, and if any arbitrary sample of emeralds should turn out to have a certain crystalline structure, it would be reasonable to infer that all emeralds have this structure.

5

Essentialism in the Social Sciences

5.1 SCIENTIFIC ESSENTIALISM AND SOCIAL THEORY

Social scientists tend to be strongly anti-essentialist. They associate essentialism with just about everything that is bad in social theory and practice – for example, with racism, social Darwinism, sexism, and other positions that play down the roles of culture, circumstances, education, or oppression in the formation of character. Those who are called "essentialists" in the social sciences are people who stress the importance of biological and genetic factors in explaining our attitudes, capacities, and so on. But scientific essentialism has nothing much to do with such attitudes, and is quite neutral on the issue of nature versus nurture.

Scientific essentialism is a theory about the sources of power in the world. If you are a scientific essentialist, then you must believe that the laws of nature are grounded in the properties and structures of things. They are intrinsic to things in the world, not imposed on them by God or anything else. You will also believe that things belonging to natural kinds must behave as they do because this is how they are essentially. A copper atom must behave as copper atoms do because it has the nuclear and electron structure that any copper atom must have, and so must display the causal powers, capacities, and so on that all copper atoms must have in virtue of being atoms of this kind. This is what scientific essentialism is all about. It is not about whether each thing in the world has within itself the seeds of its own future evolution or development. That is a much more radical doctrine, and one that I firmly reject.

Nor does scientific essentialism imply that the attitudes or capacities of people, and the members of other biological kinds, are wholly determined by their innate properties and structures. For human beings can obviously

177

acquire and lose behavioral dispositions, and acquire or lose various powers and capacities, depending on their upbringing, training, circumstances, and so on. They are not, therefore, like copper atoms. They can learn to do things, improve their skills, acquire attitudes, become indifferent to things, lose their capacities, or go out of their minds, So the idea that human beings belong to fixed natural kinds, of the sort that a copper atom might reasonably be supposed to be, is absurd.

Nevertheless, there could well be laws involving human beings. For there could be laws that hold because of properties that human beings do have just because they are human, or just because they are human beings of some variety — that is, properties that they have essentially as members of the species or as members of some variety of the species. That there are such laws cannot be ruled out *a priori*.[1] Scientific essentialism is therefore compatible with some degree of genetic determinism. Nevertheless, it is not to be understood as a doctrine that would add significantly to the nature–nurture debate. For scientific essentialism is also compatible with there being a high degree of social determinism in the development of people's capacities and in the formation of their attitudes and character. If human beings are members of a natural kind cluster (as argued in Section 4.7), then they are members of a cluster of highly variable natural kinds (as defined in Section 1.2).

My claim is that the social sciences are unlike the physical sciences in at least one very important respect: There are no laws specific to any of the social sciences that have either the status or the modality of the laws of nature. This is so, I claim, because the social sciences, except insofar as they are concerned with human beings, are not concerned with natural kinds. Human laws, institutions, social structures, cultures, political organizations, and so on are not members of natural kinds. But, according to scientific essentialism, all causal laws are grounded in the essential dispositional properties of natural kinds. Consequently, there can be no causal laws of social behavior or development that are specific to the kinds of things that are the subject of social theory.

There is, however, one social science that would appear to provide a clear counter-example to this thesis — economics. For many would argue that economics is really much more like physics than like history, sociology, or politics. Moreover, they would say that there are laws of economics which have the just same kind of necessity about them as the laws of physics. It is true that the general equilibrium theories of macroeconomics do not always yield true predictions about the behavior of real economies. But then the real economies to which these theories are applied are not ideally free market economies, and the competition between firms in these economies is imperfect. Given an ideally free market economy with perfect competition, these theories must apply and yield accurate predictions.

It may be conceded at once that economic theory is not really very much like history. For it is the theory behind the history of economies, not the history itself. Economic theory is concerned with more general issues, and stands to economic history in much the same way that evolutionary theory stands to evolutionary history, or social theory to social history, or physical and chemical theory to geological history. It is not a history, but a theory of the kind that is required to construct a history.

5.2 HISTORICAL EXPLANATIONS

There is nevertheless a puzzle about histories. Geological and social history are two branches of history. Both are concerned with causal explanation – with explaining how particular historical events came about. Yet they seem to be very different from each other. Geological history is considered to be realistic and objective. Social history is generally considered to be neither. Why should this be so? The geological formations that now exist have come about as consequences of complex causal processes involving powerful subterranean and atmospheric forces, acting on substances that are complex mixtures of chemicals – often ones that are beyond our capacity to analyze. Yet hardly anyone doubts that there is a true story to be told, and that an objective geological history of a region or a continent is possible. Not so with social history. Hardly anyone believes in even the possibility of an objective social history. If this is true, then it certainly calls for explanation.

The point has often been made that there do not appear to be any laws that are specific to any of the social sciences. There are laws of physics, chemistry, microbiology, and genetics, and perhaps a few other sciences, but there are no accepted laws of politics or society. The one social science that appears to have laws is economics. But, as we shall see, the status of the so-called "laws of economics" is dubious. In any case, the fact, if it is a fact, that there are no proper laws of society or politics does not seem to be sufficient to explain the apparent difference between geological and social history.

The social sciences are concerned with social structures and organizations, and these are obviously human constructs. They do not belong to natural kinds of the sort characterized by real essences, and the processes in which they are involved, qua things of this sort, are not processes that belong to natural kinds. The laws of social systems, insofar as there are any such laws, must therefore have a status different from the causal laws of physics and chemistry. If they are necessary, then this must be because they are analytic, or true by definition. They cannot be really or metaphysically necessary, as I shall argue the laws of nature are. Mostly, the generalizations employed by historians, anthropologists, sociologists, and other social scientists are not analytic, however. They have only the status of empirical generalizations. The traditional empiricist theory of the nature of scientific laws may thus be more or less right for the social sciences.

However, this is not what makes the difference between social and geological history. For there are no geological formations or processes that belong to natural kinds either. They pass few, if any, of the tests of natural kindhood. Consequently, there can be no causal laws specifically concerned with the earth's history. Geological formations were not made by human beings, as our political and social institutions certainly have been. But why should that be relevant? There are laws of physics and chemistry that are relevant to the plausibility of geological explanations of historical events in the earth's evolution. But there are also such laws in the background to social historical explanations. Human agents have to act in ways that are physically possible, and acceptable historical explanations will not readily attribute extraordinary powers to people.

It would be widely agreed that a historian can get things wrong. For there are certain facts of history with which any historical account must be consistent. But this is not the point that is being made by those who claim that history is not objective. The point is rather that social history has different aims from geological history. Social historians have to interpret the evidence, to make sense of it to readers whose moral and social perspectives may be very different from those of the participants in the events that are being described. Geological historians are under no such constraints. A list of the undisputed facts about the past that have to be accounted for in any reasonable history is not itself a history. A historian must interpret the facts, saying what happened, not in some colorless morally or socially neutral way, but in a living language with all of its moral and socially relative overtones. To list only what is undisputable about the past (or assign epistemic probabilities to it) would be to leave the real story untold.

The story that a historian has to tell is normally far removed from the indisputable facts. It is just one of a number of interpretations that is circumscribed by these facts, as they are interpreted within the framework of the account being given. It is not, nor is it likely to be, the only reasonable interpretation that is consistent with the facts, as they can reasonably be construed. Therefore, any claim of a historian to have got it conclusively right must be treated with scepticism. Objectivity of the kind that is to be expected in geological history may therefore be unattainable.

But what cannot be achieved in practice might nevertheless be worth aiming for. For one might try to be as objective as possible in answering the question, "What really happened?", even though no definitive answer is possible, either for lack of evidence, or lack of a neutral standpoint. And good historians do just that. In practice, "the view from nowhere" for which a historian may strive probably cannot be achieved. But one can respond sensitively to criticism of personal, social, or cultural bias if the criticism is specifically backed up by appropriate evidence. A historical account of events is not invalidated just by the truism that it is impossible to abstract from one's own personal, social, or cultural context to view what happened impartially. For such a criticism, if it were sound, would immediately invalidate all historical accounts. To invalidate a historical account on grounds of bias, one would have to show how the bias has led the historian to ignore, overstress, underplay, or distort some of the facts. A good historian naturally seeks to write history that is not open to any such refutation, and so aims to be impartial.[2]

Nevertheless, it may not be possible to be completely impartial as a historian, and even if it were, the incompleteness of the historical record, and the necessary under-determination of historical narrative by evidence, would make objectivity in history an illusory goal. So historians are probably right to insist that there is no such thing as objective history. They can aim to be impartial. They can try as best they can to understand the historical context within which the events they are describing occurred. They can take full account of the known evidence and the manner in which the evidence was gathered and preserved. But in the end they may have to admit that the account they finally give is likely to be affected by one or other of the various filters operating in this process, and by the unrecognized personal, social, or cultural biases of those involved.

If the equivalent of scientific realism in the field of history is historical realism, and historical realism involves the belief that the best historical narratives give objectively true accounts of the events they describe, then historical realism, so understood, is untenable. However, I do not much like this conception of historical realism because it confuses the epistemic problem of objectivity in history with the ontological problem of truth and realism in history. There is a real past, even if it is impossible to give an objective account of it.

Social scientists have often assumed that all sciences are like the social sciences, and hence that all knowledge is like the kind of knowledge that historians and sociologists have. Consequently, they have argued, our view of reality is colored by our personal, social, and cultural perspectives. Moreover, since there is no view of reality that is entirely objective, and uniquely to be preferred to any other, there is no wholly objective truth. There are just different perspectives on the world. The point is sometimes expressed, paradoxically, by saying that there is no objective reality. But strictly, this is a much more radical claim. It is not just the *epistemological* thesis that objective knowledge is impossible; it is a highly *metaphysical* claim that entails a kind of idealism. The leap from "there is no objective knowledge" to "there is no objective reality" is considerable, and needs independent justification. For relativism, understood as an epistemological thesis, does not entail idealism.

Even if they are right about history, and the social sciences generally, those social scientists and philosophers of social science who defend relativism are quite wrong to suppose that all sciences are really like the social sciences, or that all knowledge is like that possessed by historians and sociologists. For the physical sciences are clearly concerned with natural kinds of objects and processes, concerning which objective knowledge is undoubtedly possible. Indeed, it is not only possible, it exists in great quantity. Consequently, the epistemological argument for idealism, which has its roots in social and linguistic theory, cannot get off the ground in the physical sciences. There is a way the world is, and there are ways in which things happen in the world, and the physical sciences are able to tell us about them. The reality described by the physical sciences is not a social construction.

5.3 THE STRUCTURE OF REALIST EXPLANATIONS IN SCIENCE

Realistic explanations in science are offered as descriptions of the causal processes that result in what is to be explained. Such explanations are often layered; for some processes are naturally seen as being more basic than others. The more basic processes are described in general theories that have wide ranges of application. These general theories describe what are seen as the root causes of the phenomena, and explain something of their overall structure. But the phenomena to be explained rarely accord precisely with the patterns of events that would be predicted if the root causes were the only ones that were effective. In most cases, the observed phenomena are much more complex, and this is taken to show that many more causes are operating than just the most basic ones.

It is therefore usual in the philosophy of science to distinguish between the basic structure of the causal influences in a given field of study and the more superficial influences that may impinge on, and affect, the phenomena to be explained. Every general theory that proceeds from an idealization of the events to be explained, or of the mechanism by which they are produced, is a theory that depends on this distinction. For the sole rationale for making this distinction is to isolate the core phenomena so that they may be theorized about without reference to the extraneous forces that are likely to be significant in ordinary circumstances.

The root/subsidiary cause distinction is also reflected in Imre Lakatos's methodology of scientific research programs (Lakatos, 1970). The "hard core" of a scientific research program defines the basic structure of the events that underlie the range of phenomena to be explained. The "protective belt" consists of a number of subsidiary theories or hypotheses, developed to explain anomalies arising from the core theory, or to extend the range of the basic theory to include explanations of effects that cannot be explained satisfactorily on the basis of the hard core assumptions alone. The "dark matter" hypothesis, for example, is one that has been proposed to deal with a certain anomaly in the Relativistic Hot Big Bang theory of the origin of the universe, and so belongs to the protective belt that surrounds the core assumptions of this theory.[3] Van der Waals' equations of state are laws that belong to the protective belt of the kinetic theory of gases. For the theory in which they were derived is an extension, or modification, of the core kinetic theory. In this case, the extended theory was designed to deal with forces that were known to be operating, but were ignored, in the original model.

Although this pattern of explanation and this kind of structuring of scientific research programs are accepted as normal and appropriate in the physical sciences, they are often regarded with suspicion in the human sciences. It is true that there are some respectable theories in these sciences that aim to explain just the basic structure of causation in their fields. Theoretical linguistics, for example, aims to develop a general theory of language structure and acquisition that is applicable to all human languages. Logic aims to develop a general theory of rationality adequate to explain the basic structure and dynamics of rational belief systems (Ellis,

1979). Neoclassical economics aims to develop a general theory of the structure and dynamics of market economies. Within each of these disciplines, there are research programs characterized by hard-core assumptions and protective belts of hypotheses. But abstract model theories such as those in theoretical linguistics, logic, and economics are rare in the human sciences, and for the most part they are held to be inappropriate. A "hard core" set of assumptions in sociology, history, or politics, for example, would be considered ideological and therefore contrary to the ideal of objectivity in which practitioners of these disciplines implicitly believe (whether or not they think it is achievable).

The question therefore arises: When are the patterns of explanation and theory development that are characteristic of the physical sciences appropriate in a given area of study, and when are they inappropriate? When is it reasonable to assume that a general theory of causation in an area of study is possible? The answer that will be given here is that it is reasonable to believe that a general theory of causation in an area is possible only if the kinds of entities under investigation can reasonably be assumed to belong to natural kinds, and the causal processes involved in their actions and interactions can reasonably be supposed to be displays of the intrinsic causal powers, capacities, liabilities, and so on of things of these kinds.

5.4 ECONOMIC ESSENTIALISM

The social scientific theory that has most often been considered to be an objective science, and whose theorems might be said to be necessarily true, is neoclassical economics. It is widely believed, for example, that the neoclassical equilibrium postulate is necessary, and has the status of a law of nature. For the postulate is thought to describe what must of necessity be the case in any pure competitive market economy in equilibrium, and therefore to describe the state toward which any such economy must tend, if it is left to its own devices. Let us call this theory about the nature of economics "economic essentialism."

Economic essentialists believe that there is a positive science of economics, which is comparable in status to physics, and that its laws have the same kind of necessity. It is a science, they would say, that differs from physics in subject matter and in the kinds of observations that can be made. But it has a body of established high-level theory, just as physics has, a methodology that is similar, and a number of major research programs under way. The methodology that is usually deemed to be appropriate to economics is Lakatosian (Lakatos, 1970). Consequently, economists now distinguish, as Lakatos did for research programs in the physical sciences, between the "hard core" assumptions, which define the programs and are held dogmatically, and the theories or hypotheses belonging to their protective belts, which are more open to empirical refutation. However, economics is not fundamentally like physics – or so I shall argue. There are important ontological differences between the kinds of entities and processes studied in these areas that make the comparison weak. The models of market economies developed by economists do not have the status of physical scientific models, as the comparison suggests. Nor is Lakatos's methodology of scientific research programs one that is wholly appropriate for economics. For there is no principled way of making a satisfactory distinction between economic hypotheses that should, or should not, belong to the hard core. There is, for example, currently a dispute about which hypotheses belong to which group in the neo–Walrasian research program (Backhouse, 1993). The belief that these two fields of study are basically similar is therefore wrong-headed. It has also done a great deal of harm: For it has comforted those who endorse *a priorism* in economic theory, and reinforced the all too prevalent attitude of indifference to predictive failure in economics.

A. Rosenberg (1992) has argued on historical grounds that economic theory is not a body of empirical scientific knowledge about the underlying structures of real economies, but is, rather, a mathematical theory developed mainly *a piori* with little concern for the facts about people or the economies to which the theory may be applied. I think he is basically right about this: The development of economic theory, and the acceptability of its models, do not sufficiently depend on how well they can explain the facts. As it is practiced, economics is not responsive enough to evidence to be considered an empirical science, comparable to, say, physics or astronomy, despite the belief of many economic theorists that their approach is a scientific one. It will be argued here that the kinds of processes that are the concern of economists are ontologically different from the kinds of processes studied in physics. They are not natural kinds of processes, and consequently do not have any real essences for the "hard core" assumptions of a research program to speculate about. There is, therefore, no *de re* essence/accident distinction in economics that could possibly provide an ontological basis for the "hard core/protective belt" distinction of Lakatosian methodology. Economic entities and processes might indeed have nominal essences – that is, be linguistically defined. But if the con-

185

ventions of language do not reflect underlying natural distinctions, then any laws concerning these entities or processes can only be trivial – that is, true just by definition or convention.

If this is right, then there is need for much more humility in economics – a greater willingness to let the detailed facts about our economies be our guide in theory construction, rather than, say, the claims of the supposed "hard core" of the neo-Walrasian research program. I would therefore advocate something more like old-fashioned Baconian empiricism as the appropriate methodology for economics. One can perhaps modify the culture of a society in order to change people's behavior to fit one's theories better – which was the agenda of the "Old Left" and is now the agenda of the "New Right." But if economics is to become a positive science, capable of predicting outcomes, and prescribing policies for achieving given outcomes, in a given unreconstructed society, then it must abandon its dogmatism about its basic assumptions, and adopt a more open empiricist stance. Bacon once said, "we cannot command nature, except by obeying her" (Farrington, 1951, 7). The same is not true of economies. For they can be – and are – commanded. But if we want our economic advice to be compatible with our culture and way of life, then we had better observe carefully the culture and way of life we wish to preserve.

5.5 ECONOMIC MODELS

There are two principal models of the market economy – the neoclassical and the Keynesian. Neither is entirely satisfactory as an economic model. J. Tobin (1977) has argued that the evidence would seem to support Keynesianism rather than neoclassicism. However, neoclassical economics is the most widely accepted economic theory, and the most highly developed. The neoclassical theory proposes a model of the economy based on a general equilibrium postulate, according to which aggregate demand in any given sector of a free and perfectly competitive market must always be equal to the gross production of goods and services in that sector.[4]

The question I am concerned with here is: What is the status of a model such as this? It is clearly a theoretical idealization of some kind, since the markets of ordinary economies are neither free nor perfectly competitive. But, as we shall see, it is not like most of the theoretical ideals in physics or chemistry, which are intended to capture the real essences of natural kinds of processes. Nor is it just a limiting case that is of mathematical or theoretical interest, but is not intended to be directly relevant to the un-

186

derstanding or management of a real economy. For the model is used, and is applied directly to real economies, as though they could be accurately represented by it.

I argue here that the neoclassical model, which is often thought to be a scientific theory comparable to, say, classical thermodynamics,[5] is just a quasi-analytic system that can be applied more or less fruitfully to interpret, and to justify policies for the management of, market economies. But it can make no claim to be descriptive of the processes essentially involved in the workings of such economies.

In the sciences that are concerned with natural kinds of objects and processes, there is always a basic underlying structure to be described. For objects and processes belonging to natural kinds are objectively and sharply distinguished by their intrinsic natures. Each member of a natural kind has certain intrinsic properties and structures in virtue of which it is a member of that kind. These are its essential properties and structures. In chemistry, the essential properties and structures of the chemical elements and compounds and of the processes in which they may be involved are described in detail. The theories we have about these properties, structures, and processes are of course intended to be realistic. Consequently, the correct philosophical position to hold with respect to chemical theories is realism. The models of atoms and molecules and of the processes of chemical combination in which they may be involved are to be taken with ontological seriousness as descriptive of the underlying reality.

But the economy of a country is not an object that belongs to a natural kind, and the processes of the market are not instances of natural kinds of processes.[6] The economy is a socially constructed system, heavily dependent on human conventions, and different economies have different mechanisms for achieving their objectives. Even market economies differ from each other in various ways, both structurally and in how they work (Hutton, 1996; Fukuyama, 1995). Consequently, we should not expect to find a unique and simple underlying reality for our economic theories to describe.

5.6 NEOCLASSICAL *A PRIORISM*

There are some who believe that the neoclassical equilibrium postulate is necessarily true, and has the status of a law of nature. That is, they think that the model describes what must of necessity be the case in any market economy. Specifically, it tells us how any pure market economy would have to function under conditions of perfect competition if it were not influenced by any external forces. Call this "the essentialist theory."

187

The founders of neoclassical economics – W. Jevons, L. Walras, F. Edgeworth, I. Fisher, and V. Pareto – all took this position. For they believed that economics could be developed as a pure science. Geometry and mechanics were their models. They did not take economics to be an experimental or inductive science, like chemistry or anthropology, and they saw little need to support their theories with empirical evidence. The appropriate method, they supposed, was a deductive one, and its ultimate support lay in the self-evidence of its axioms. I quote from Walras:

> The pure theory of economics is a science which resembles the physico-mathematical sciences in every respect . . . The mathematical method is not an experimental method; it is a rational method . . . The physico-mathematical sciences do go beyond experience as soon as they have drawn their type concepts from it. From real-type concepts, these sciences abstract ideal-type concepts which they define, and then on the basis of these definitions they construct *a priori* the whole framework of their theorems and proofs. After that they go back to experience not to confirm but to apply their conclusions. [*Elements of Pure Economics*, p. 71. Quoted by Brian Toohey, in *Tumbling Dice*, p. 10.]

The founders were not much concerned that their theories were abstract, and referred to idealized objects or circumstances. Many physical theories do that. There may not be any ideally rational agents or perfectly competitive industries. But then there are no point masses or ideal gases either. Therefore, neoclassical theory is not to be distinguished from physical theory on this account. The difference, they thought, is only one of subject matter, a view echoed by M. Friedman (1953, 4) and many other economists. Moreover, they supposed that the laws of economics, which describe how their theoretical models operate, have as much claim to be recognized as laws of nature as any of the laws of mechanics. The fact that these economic laws describe the behavior of idealized objects in ideal systems does not count against their being laws of nature. Many of the laws of physics, which are certainly laws of nature, do just this.

This conception of the nature of economic theory and of the status of its laws is not wholly out of date. I do not suppose there are many economists today who think of the laws of economics as comparable in every way to those of physics. But many evidently do think that the assumptions economists make in deriving the laws of market equilibrium are so obvious that no justification is required. Consequently, many economists adopt what K.R. Popper (1959, 82–4) used to call "the conventionalist stratagem." If predictions fail, then the fault does not lie in the theorems, but in their application. Thus, if real economies do not behave as they should according to the theory, then either (a) the true market equilibrium level has not yet been reached, or (b) there are imperfections in the market responsible for the failure, or (c) people are not acting rationally – excuses that are in fact both always available and always supportable.[7]

It will be argued here that the conception of economic theory that feeds such dogmatism is fundamentally wrong. Economic theories do not contain theoretical ideal-type concepts of the kinds that are to be found in the physical sciences, and the laws of economics are not physically necessary, as the laws of nature are.

5.7 THEORETICAL IDEALS AND EXPLANATORY FRAMEWORKS

The ideals of physical theory are of two kinds. First, there are those that allow us to describe how various natural kinds of systems would behave independently of sundry and contingent causes. The concept of a closed and isolated system, for example, is an ideal of this kind. Such ideals allow us to focus on the natural kinds of systems or processes that are isolated theoretically in this way, and to describe their essential natures. But although there are ideals a bit like this in economic theory – for example, the ideal of a closed economy – there are no natural kinds of economic systems or processes to be theoretically isolated. The class of economies known as "market economies" is a collection of economies with a more or less similar structure. But it is not a natural kind. Consequently, economic theorists can never hope to discover laws of nature governing closed market economies.

Second, there are theoretical ideals that set up explanatory frameworks and provide the necessary framework principles. These ideals form the bases for theoretical explanations. The explanations they provide are two-stage explanations. There is, first, the explanation of how the system would behave if it were ideal. This is the *abstract theoretical* part. Then there is the explanation of the difference between the actual and the theoretical behavior of the system. This difference is the effect that remains, and for which we must now find sufficient cause. The explanation of this difference is the *causal explanatory* part. If the models of economic theory are theoretical ideals, then they have to be judged as theoretical ideals of this kind. The test of such an ideal is the adequacy of the theoretical framework it provides. Are the laws that are supposed to govern the objects in the model theoretically derivable from any more fundamental laws that we are independently justified in accepting? If so, then the model has a sound theoretical basis. Can precise causal explanations be found for the effect that remains to be explained – that is, for the difference between the actual and the theoretical behaviour of the system? If so, then the model is empirically adequate. Does the theoretical framework of neoclassical economics pass either of these tests? I think not.

A theoretical framework can be explanatory in economics, or in any other field, only if it is empirically adequate. In the field of economics, there is no more fundamental theory – for example, of human nature, or of society – from which the supposed economic principles can be derived – at least, none that we have any other reason to accept – and few economists nowadays would think otherwise.[8] Therefore, it is just the empirical adequacy of the theoretical framework that matters. The question then is whether the failure of any real market economy to behave as it should according to the theory can be adequately explained by its failure to satisfy the requirements on the model. I do not think so. It is not enough to gesture in the direction of market imperfections, or disequilibria, or human irrationality to explain the failure. A decent quantitative causal explanation of the difference between the theoretical behavior of the ideal system, and the actual behavior of the real system, has to be given, if the theory is to pass the test of empirical adequacy. It is the considered view of many of the world's leading economists that no economic model yet proposed passes this test.[9] There may indeed be no single model that can provide a theoretical framework that is empirically adequate for all market economies. The dynamics of different market economies could be significantly different from each other, depending perhaps on the social structures within which they exist (Fukuyama, 1995).

5.8 THEORETICAL MODELS IN PHYSICS AND ECONOMICS

The language of economics certainly looks very like the language of many physical theories. There are concepts of perfect competition, free markets, ideally rational agents, and the like. One is tempted therefore to suppose that economic theories are just like certain idealized physical theories – theories concerning such things as black body radiators, ideal incompressible fluids, and so on. Indeed, the claim that economic theories are closely analogous to such physical theories is often made, when doubts are raised about the scientific credentials of economic theories. But if the analogy held, then the laws applying to the various economic ideals should turn out to have the same status as those applying to some of these physical ideals. That is, they should turn out to be laws of nature.

Consider first the concept of perfect competition. Perfect competition, by definition, exists within an industry when and only when:

1. There are so many producers of the goods or services supplied by the industry, and so many buyers of these goods or services, that no one producer or purchaser contributes enough to the overall supply of, or demand for, these goods or services to alter the sale price.
2. There are no price-fixing arrangements between producers, and no collaborative buying arrangements among consumers, capable of affecting the sale price.
3. There are no obstacles to individuals or firms entering or leaving the industry.
4. There is no brand loyalty or product differentiation, and purchasers will, without discrimination, buy at the right price from any supplier.

The concept of a perfectly competitive industry is certainly a limiting-case concept. It is derived by extrapolating to zero the effects on price of factors other than the costs necessarily involved in the production of the goods or services. Transaction costs are commonly set to zero.

Transaction costs are costs associated with economic transactions other than the direct costs of the purchases or the charges for goods or services provided. According to this definition, transaction costs include such varied items as:

1. Execution costs – brokerage, stamp duty, license fees, tariffs and sales taxes.
2. The cost of product and job search, when expressed in terms of the opportunity cost of time.
3. The delay in the completion of a tatonnement process – for example, the matching of a buy order by a sell order in the financial markets.

Kim Sawyer argues that economic theorists commonly assume transaction costs to be negligible or even non-existent in much more specific models than the general one. No transaction costs are embedded in the equilibrium asset pricing models of finance (for example, the capital asset pricing model and arbitrage pricing theory) and no transaction costs are present in the standard Black-Scholes option pricing model. Since the turnover in financial markets is sensitive to transaction costs (for example, the Taiwanese market, which has the lowest brokerage in Asia, also dominates in terms of turnover), the assumption of zero transaction costs is non-trivial. Yet, typically, theoretical models suppress transaction costs. One reason presumably is that the most elegant and tractable theoretical results are obtained in models without transaction costs. It is also the case, however, that many economic purists regard the functioning of markets as tantamount to a physical process, and transaction costs as an impediment to the smoothness of that process. Unsurprisingly, taxes and other transaction costs are commonly termed "market frictions," and their minimization is equivalent to oiling the process.

Clearly, perfect competition is supposed to be more than just a limiting case, since the model is constructed to analyze the effects of a certain limited range of factors on prices. And this is already enough to make it a legitimate theoretical model. The question I want to ask is: How like or unlike is the model of perfect competition to that of those physical ideals concerning which laws of nature may be formulated – for example, the ideal of a perfectly black body? Prima facie, the two models are of the same kind. One is constructed by ignoring, or setting to zero, those factors other than the costs of production that may affect the prices of the goods or services sold in a given industry; the other is constructed by ignoring or setting to zero those factors other than temperature that may affect the pattern of radiation emitted or absorbed by a body. Nevertheless, there are important differences between them. The heat theory model is a "natural kind" model. It is true that there are no perfectly black bodies in nature. But there is such a thing as temperature radiation; and this is a natural kind of process.[10] We know it is a natural kind of process because it is intrinsically different from any other kind of process, and is the display of a universal tendency of matter. Accordingly, the process is governed by some precise and distinctive universal laws. Moreover, these laws are not trivial. They do not, for example, follow from the definition of a perfectly black body. They are quantitative laws that hold exactly of all and only black body radiators. Furthermore, they are theoretically connected to other, more fundamental, laws concerning Planck's quantum of action, and so integrated with the rest of physics. The laws of black body radiation thus have a sound theoretical basis.

Nothing comparable is true of perfectly competitive industries. A perfectly competitive industry is not a natural kind of object, and there are no natural kinds of processes to which such industries would give rise. There are no laws governing the behavior of prices for goods or services in perfectly competitive industries other than those that are immediately derivable from the defining properties of perfect competition. If you want to increase profits in a perfectly competitive industry, you either have to produce the goods or services you wish to sell more efficiently, or improve their quality, or the services you provide in marketing them. But this is not a law of nature. If, by definition, a bachelor is an unmarried man, then it

192

is not a law of nature that there are no married bachelors: It is a truism. Likewise, it is not a law of nature, but a truism, that business people intent on increasing profits in a perfectly competitive industry in equilibrium have to increase efficiency, or improve their quality, or their marketing. They cannot increase prices, since, by definition, they are price takers. They cannot purchase their labor or raw materials more cheaply, since these prices are already fixed by the market. They cannot increase production, because the marginal cost of producing an extra unit makes this uneconomical. The only thing they can do is use the same labor, materials, and equipment more efficiently in making or selling the product.

The laws of black body radiation, in contrast, do not follow from the definition of a perfectly black body. A perfectly black body is one that absorbs all of the radiation which falls upon it. But Planck's radiation law does not follow just from this definition. The definition manifestly does not imply that the total quantity of energy $E_l dl$ per second from a unit area of the surface of a perfectly black body in the wavelength range l to $l + dl$ at the absolute temperature T is given by

$$E_{l}.dl = jl^{-5}/(e^{k/lT} - 1).dl$$

(where j and k are constants). But this is what Planck's radiation law states.

The models of economic theory therefore do not have the same kind of role as the models of physical theory. Economic models are fundamentally different from the theoretical models of physics because they are not essentialist models. They do not describe quantitatively any processes of a kind that must occur in any real market economy. The processes they do describe quantitatively are just those that must, by definition, occur in an ideal market economy, but that need not occur in any real economy. Nor are there any laws relating the variables that describe an ideal market economy, apart from those that follow from the definitions employed and from some more or less crude generalizations based on observations of existing "market" economies.

There are, of course, quantitative relationships between defined quantities, which are descriptive of an economy, and that must hold true in any economy to which these definitions are applicable. But these quantitative relationships hold by definition; they are not observational or experimental truths and they do not stand in need of theoretical explanation or justification. There are also various *ceteris paribus* laws that follow directly from these definitional truths. For example, if a quantity Q is equal by definition to A + B + C − D, and A, B, C, and D are all manipulable variables, then it is true that other things being equal, if you decrease D, you will increase Q. It does not follow, however, that in any real economy, if you decrease D you will increase Q. For this to follow, it needs also to be the case that the variables A, B, C, and D can be varied independently of each other in the given economy. And this may not be the case. In real economies, there are often high degrees of interdependence of economic state variables, and varying any one of them can have flow-on effects to some or all of the others. Moreover, the nature of this flow-on can vary considerably from one economy to another, depending on the culture or expectations of the people or the firms that make up the economy.

5.9 METHODOLOGY

If the pure market economies of neoclassical economics are not natural kind models, then what are they? In my view, they are just mathematical constructs that describe idealized economies that behave, as perfectly competitive free market economies must behave, according to neoclassical theory. Consequently, their hypotheses are all either analytically or trivially true of the model economies, however plausible as assumptions about real economies they may be. They are useful, and useful only, to the extent that real market economies behave in ways that approximate to these ideals. But they do not describe the essential nature of real market economies, or of the market forces that operate in them, as neoclassical essentialism supposes. For market economies and market forces have no essential natures. Market forces are different, and behave differently, in different economies. They also behave differently in different sectors of the same economy, at different stages of its development, and at different times in the business cycle. In other words, the idea that there are laws of operation of market forces comparable to those of, say, gravitational forces, is nonsense. Yet, by endorsing Lakatosian methodology, many economists have embraced a methodology for their discipline that presupposes essentialism. For, in doing so, they have assumed without argument that the methodology appropriate to economics is that appropriate to a discipline such as physics that is primarily concerned with the description of natural kinds of objects and processes. Disputes about the status of neoclassical economics as a science have consequently centered on the question of whether the research program it defines is progressive.[11] However, in an area that is not concerned with the description of such processes, the embrace of Lakatosian methodology can easily be just an excuse for dogmatism. As there is no principled way of making the

194

hard core/protective belt distinction in such an area, the decision to hold fast to certain assumptions should be much more open to revision. Lakatos' methodology is not appropriate for economics since it cannot be assumed that there is a unique underlying mechanism at work driving real economies that it is the business of the hard core of the research program to describe.

Lakatos himself is largely to blame for this since he gives the impression that the "hard core"/"protective belt" distinction on which his theory is founded is sociological in origin and justification. Scientists, in their wisdom, hold some propositions (those of the hard core) dogmatically, and refuse to countenance any objections to them, for no other reason, apparently, than that they are basic to their theory. Objectors only show themselves not to be bona fide researchers in the field. But dogmatism about some propositions rather than others can be justified in a science only if there is a significant distinction between these propositions and the other hypotheses of the theory. If, for example, the propositions of the hard core are basic postulates about the essential natures of the objects or processes involved in the area under investigation, then there is obviously some point in protecting these from refutation by observations of phenomena that can plausibly be explained as due to causes that are acting independently of these essential properties or structures. In short, a two-level methodology of scientific research programs of the kind described by Lakatos may be appropriate if there is an ontological distinction between basic structure and perturbing influences. But it may well be quite inappropriate, as it is in economics, if the objects or processes we are dealing with have no essential properties or structures.

Until early in the twentieth century, space was assumed to be an independently existing entity having its own essential nature. This was thought to be correctly described by Euclidean geometry. But there were problems. The measured velocity of light did not vary as it should, according to the theory of electromagnetism then accepted, with the inertial motion of the observer. Indeed, the same equations (Maxwell's equations) seemed to apply independently of the inertial systems to which they were referred. But for this to be possible, the transformation equations between inertial systems had to be more complicated than the simple Galilean transformation equations of the Newtonian world. The required transformation equations were the Lorentzian ones, which seemed to imply that space and time were not independently existing entities, but dependent on each other as aspects of a more comprehensive entity which included both space and time. However, the ontological shift did not come at once.

Lorentz and Fitzgerald put forward a length contraction hypothesis designed to preserve both the Euclideanness of space and its independent existence. The decisive ontological change, which replaced Euclidean space with Minkowsian space-time, did not occur until after the publication of the Special Theory of Relativity in 1905.[12]

When the concept of space as an independently existing entity was abandoned in 1905, Euclidean geometry could no longer enjoy the privileged status it had enjoyed in the nineteenth century. As Reichenbach (1958, 30–37) argued, one could, dogmatically, hold on to belief in a Euclidean space, and make changes in the physics (for example, by introducing universal forces) to accommodate this belief. But this would be arbitrary. What counts, Reichenbach argued, is the overall simplicity of the theory, including not only the geometry, but also the physics. For geometry must now be seen to be part of physics, rather than something presupposed by physics.

Similarly, it seems to me, there is no case for working with an economic theory that allows just the neoclassical model to define the framework for the subject. For that is to treat the model as though it were essential to the nature of any real market economy. However, there is no reason to think that this is so. Therefore there is no good reason to protect the hypotheses of the neoclassical model from refutation by surrounding them with a protective belt, as Lakatosian methodology recommends. To do so is not good scientific practice: It is dogmatism. It is like insisting that the geometry of space must be Euclidean, whatever the empirical evidence may be. What matters in economics is its empirical adequacy as a theory, and its overall simplicity. Whether the best theory from this empirical standpoint includes any or all of the neoclassical equilibrium hypotheses is an open question. It is a question of which economic theory provides the best explanation of the facts. It is not a question that can be settled *a priori,* or by reference to the theoretical commitments of economists.

Of course, if the neoclassical research program had turned out to be progressive, then there would have been good reason to persevere with it. For the very success of the program would show that my claim that market forces do not constitute a natural kind, or set of natural kinds, is dubious. For how else could one explain the success of the program, except on the basis of this assumption? I do not want to be too dogmatic about what natural kinds there are. Nevertheless, as Rosenberg (1992) argues at length, the neoclassical research program has not been progressive. So I see no good reason to continue with it as though it were the only contender in the field.

How then should we approach the study of economics? I suggest that the appropriate methodology for economics is just that of empirical sociology. That is, if you want to know how markets of various kinds behave in a market economy, then you must conduct sociological research. You should not rely on models based on *a priori* judgments about the effects of prices, or spending power. Nor should you assume that people will tend to act rationally, according to criteria for rational behavior that no one could possibly use in practice. Rather, you must find out empirically how business people set prices, what their hiring and firing policies are, what kind of work people want, or are prepared to take, what their attitudes are to foreign and local products, how people in different groups in the community spend or save their money, and all of the thousands of things that economists need to know if they are to make reasonable and informed predictions, or recommendations to governments on policies. In other words: You must be empiricists in the old, and often maligned, sense of this term. There is nothing wrong with getting your hands dirty by studying the facts about the behavior of real economies, or the people who live and work in them.

It is often said that the science of antiquity collapsed through its obsession with *a priorism*. Be that as it may, economic science is certainly in danger of losing much of its remaining credibility because of this obsession. Economic *a priorism* has already led the world into one depression of major proportions. It is now in the process of destroying the social contracts that exist in most Western industrialized countries, as governments everywhere lower taxation levels to reduce the perceived burden of government on business enterprises, and then struggle to reduce government expenditures to match the falling revenues. If economics is to retain credibility, it must seek a more empirical basis for its theorizing.[13] Obviously, it cannot establish a new research methodology overnight. To construct a satisfactory empirically based economic model for a country must be considered a long-term project. Moreover, any model so constructed would need to be continually updated. There would need to be an ongoing research program to monitor the variables, and the statistical correlations between these variables, as information technology, international relations, and the expectations of people in society, change. However, one could, in the meantime, fall back on Keynesianism, which, for all its faults,[14] is not an *a priori* system,[15] and did serve the world very well for thirty years or so following the outbreak of the Second World War.[16]

When I speak of empiricism in economics, it is not the kind of empiricism advocated by Friedman (1953), who viewed economic theories merely as instruments, in the manner of the positivists. For Friedman's preferred methodology was still Lakatosian, even though he adopted this positivist (and hence empiricist) economic metatheory. My objection is to the use of this methodology in an area in which we cannot reasonably suppose that there are any underlying natural processes whose natures we are seeking to reveal. The empiricism I should advocate is not instrumentalism, which is meta-theoretic, but a methodological empiricism – an investigative method which relies heavily on data collection and analysis. What we should be opposed to is the kind of abstract theorizing that Francis Bacon railed against in his famous attack on Scholasticism.

> ... the wit and mind of man, if it work upon matter, which is the contemplation of the creatures of God, worketh according to the stuff, and is limited thereby; but if it work upon itself, as the spider worketh his web, then it is endless, and brings forth indeed cobwebs of learning, admirable for the fineness of thread and work, but of no substance or profit (Farrington, 1951, 59).

Bacon's inductive methodology may not be wholly inappropriate for economics.

1. It is my belief (Ellis, 1979), for example, that we have an in-built, and so presumably genetically encoded, epistemic value system that is the source of our common ideals of rationality.
2. K. Windschuttle (1996) gives an excellent account of how and how not to write history. I am indebted to him for his discussion of the issue of objectivity in history. C. Behan McCullagh's (1998) volume also provides an excellent defense of objectivity in history.
3. The observable quantity of matter in the universe is very much less than it should be according to the theory.
4. There are of course many variants of the neo-classical model. But since what we have to say is applicable to them all, we will speak as though there were only one.
5. This comparison is explicitly drawn and discussed at length in Nelson (1992).
6. The point that the subject matter of economics is not naturally structured into kinds is carefully argued in Nelson (1990). It is also suggested in Rosenberg (1992), although the argument is not systematically developed there. "... the real source of trouble," he says, "for the attempt to find *improvable* laws of economic behaviour is something that has only become clear in the philosophy of psychology's attempts to understand the intentional variables of common sense and cognitive psychology. . . . 'beliefs' and 'desires' – the terms with which ordinary thought and the

social sciences describe the causes and effects of human action – do not describe "natural kinds." They do not divide nature at the joints. They do not label types of discrete states that share the same manageably small set of causes and effects and so cannot be brought together in causal generalizations that improve on our ordinary level of prediction and control of human actions, let alone attain the sort of continuing improvement characteristic of science. We cannot expect to improve our intentional explanations of action beyond their present levels of predictive power" (p. 235).

7. The failure of the neoclassical paradigm in economics is actually one of the main sources of its strength. For every failure of the theory can be seen as strengthening the case for the neoliberal agenda of deregulation, removal of impediments to free trade, reduction of government expenditures on health, education, and welfare, and so on. Failure only adds to the conviction of governments that their "reforms" have not yet gone far enough.

8. The assumptions that we have to make concerning human choice behavior, if we wish to derive the general equilibrium postulate, are widely agreed to be bizarre.

9. Rosenberg has argued the case for this at length in Rosenberg (1992).

10. The kind of radiation known as "temperature radiation" is emitted by all substances. Its intensity increases rapidly with temperature, and peaks at shorter and shorter wavelengths. At low temperatures, such radiation is of low intensity and is in the deep infrared. Hence, it is invisible at these temperatures. At higher temperatures, this kind of radiation becomes visible, and things begin to glow red-hot. At still higher temperatures, this radiation becomes ever more intense, and its color changes from red, to yellow, to white, and even to blue, in the case of some stars. However, while this general pattern is universal, and the same for all substances, the precise nature of the radiant energy varies from substance to substance, depending on what proportion of the radiant energy falling on it it absorbs and what it reflects, at various wavelengths and temperatures. In the nineteenth century, it was established thermodynamically by R. Kirchhoff that the ratio of radiant energy emitted to radiant energy absorbed at any given temperature and wavelength must be the same, and so equal to the radiant energy emitted by a body that absorbs all of the radiant energy that falls on it – that is, a perfectly black body. The concept of a perfectly black body thus has a central role in the analysis of the pattern of temperature radiation. This is a good example of what is meant in science by a theoretical ideal. It is a theoretical object of considerable importance. All real objects emit and absorb temperature radiation. But the overall pattern of such radiation is obscured by the prevalence of other kinds of radiation, as well as by the fact that all ordinary bodies are colored, and so emit and absorb radiation selectively. To study the unique role of temperature radiation in the establishment of thermal equilibrium, it is necessary to abstract from what occurs in any real case to consider what would occur in the ideal case in which all of the incident radiant energy is absorbed by the participating objects. That is, we have to consider what would happen if the objects were perfectly black.

11. See, for example, Rosenberg (1992) and Rappaport (1995). Although Rosenberg thinks that the research program of neoclassical economics is not progressive, he concedes that agricultural economics has exhibited some substantial predictive improvement over the years. However, he argues that this is more technological

199

than economic. Rosenberg's arguments are challenged in Rappaport (1995). But he does not claim to have falsified any of Rosenberg's claims. His verdict is the Scottish one of "not proven" (Rappaport (1995, 156).

12. The conception of space-time as an independent entity having its own essential nature itself gave way (in 1916) to a more comprehensive entity – a physical world of space-time and matter/energy, the essential nature of which is described in Einstein's General Theory of Relativity. So, space-time, as we now understand it, is just an aspect of the physical universe. Its geometry is not essential to it, but is dependent on the global distribution of matter and energy.

13. Rosenberg (1992, 228) argues that "anyone with much knowledge of the history of economic theory will agree that the discipline does not seek to respond to empirical data in the way characteristic of an empirical science – even a theoretically impoverished one. If social contract theory is too Procrustean a bed for economic theory, and empirical science is too demanding a status, is there not some other interpretation of the aims and methods of economic theory that will do full justice to its scope and its insulation from data?" (passage quoted by Rappaport, 1995, 137). Rosenberg's ultimate conclusion is that economic theory is applied mathematics – that is, it has the status of an applied calculus.

14. Keynesian economic management had trouble in the 1970s with the problem of "stagflation" – simultaneously low levels of demand and high inflation. However, the inflation at the time appears to have been cost-push, rather than demand-pull, and was probably more of a political problem than a purely economic one. Therefore, stagflation might well have been just the occasion, rather than the cause, of the rejection of Keynesianism. The neoclassical revolution of the mid-seventies that overthrew Keynesian theory and reestablished the neoclassical position was in any case quite extraordinary, and I have never seen a satisfactory account of it. It was all the more extraordinary because Keynesian theory was itself a response to a crisis in neoclassical economics – the 1930s depression. If economics were a genuine empirical science, it is incredible that this should happen. For never before in the history of science has a discredited theory been revived in such circumstances and bounced back to replace the theory that successfully replaced it.

15. Tobin (1977) lists four central Keynesian theses that he claimed stood up well in the light of experience up to the time of his writing. For example, Keynes claimed that in any pure market economy, there will be people who cannot find jobs willing to work at or below prevailing real wages and who will have no effective way of signaling their availability (p. 459). The high levels of unemployment in most industrialized countries would seem to bear this out. Indeed, all four Keynesian theses discussed by Tobin are just as plausible today as they were in 1977.

16. In most Western countries, neoclassical economics was all but destroyed by the events of the Great Depression. In the period of Keynesian dominance, the Western industrialized world enjoyed unparalleled prosperity and high levels of employment and growth. Normally, a dominant scientific theory or paradigm remains unchallenged unless, or until, it runs into serious, and apparently insuperable, difficulties. But this was not so in the case of economics.

Part Four

Laws of Nature

6

Theories of Laws of Nature

6.1 INTRODUCTION

According to A.R. Hall, the idea that nature is governed by laws does not
appear to have existed in the ancient Greek, Roman, or the Far Eastern
traditions of science (Hall, 1954, p. 171). Hall suggests that the idea arose
as a result of a "peculiar interaction between the religious, philosophical
and legalistic ideas of the medieval European world."[1] I think there were
probably other sources of the idea. There was, for example, the influence
of Euclid's geometry and Archimedes' statics in the medieval period in Europe, and the attempt that was made then to apply geometrical methods
to the study of mechanics. These ancient works must have suggested to
the medievals, as geometry had suggested to the ancient Greeks, that
knowledge is structured, and the successes that were achieved in the early
medieval period in solving problems of mechanical equilibrium, making
use of such principles as the law of the lever and the principles of mo-
ments and virtual work would certainly have added substance to the idea
that nature is governed by laws. Be that as it may, it is true that the laws of
nature were conceived from medieval times as general principles of motion (in the sense of locomotion) and equilibrium – that is, as principles
governing the kinds of changes that can occur in the world and the kinds
of equilibrium states that can exist.

The modern concept of a law of nature is no longer restricted in its
application to the principles of mechanics, as the medieval concept was.
It includes the principles of all kinds of changes and states of equilibrium.
However, it is not so very different in conception. For the laws of nature
are still widely thought of as governing nature – as imposing order and
structure upon it – as if by the command of God. I shall argue that this is
quite the wrong way to think about laws of nature.

Evidently the laws of nature are not all of the same kind. Some are global laws, in the sense that they apply directly to all things – that is, to all events and processes. Others are more specific, and apply directly only to specific kinds of things or processes. So laws might be distinguished by their scope or generality. However, if laws are what they are usually represented as being, it is hard to see how we could possibly do this, because, if it is a law that all As are Bs, then everything falls within the scope of its universal quantifier, if not as an A, then at least as a non-B. For "All As are Bs" is logically equivalent to "All non-Bs are non-As." However, laws of nature are not just universal generalizations, as this formula suggests. In it simplest form, a law of nature says that for all x, if x belongs to the natural kind A, then x will be intrinsically disposed to be involved in an event of the natural kind B in circumstances of the kind C. For simplicity, we might say that all As are Bs in circumstances of the kind C. But however we might choose to express them, it is true that laws of nature involve natural kinds of events. In that case, the terms occurring in the contrapositives of laws of nature do not, unless accidentally, also refer to natural kinds of events.[2] Hence, from the perspective of an essentialist about laws of nature, there is a clear distinction to be drawn between the direct or *positive* formulation of a law and its *contrapositive* formulation. The fact that all As are Bs in circumstances of the kind C implies that all non-Bs are non-As in such circumstances does not imply that the law is one about non-Bs. If the first of the two logically equivalent propositions applies directly to the members of a natural kind, then it is not necessarily true that the other one does too. Normally, it applies only indirectly (via its contrapositive) to these very same entities. Thus we may distinguish between the positive and the contrapositive formulations of laws. The positive formulation of any law is one that applies directly to the natural kinds of objects or processes named in its antecedent.

The distinction between the positive and the contrapositive formulations of laws is related to probability theory's distinction between *reference* class and *attribute* class. In fact, for probabilistic laws, there is a clear difference between $P(B/A)$ and $P(\sim A/\sim B)$, since these two quantities are equal only in special cases. Even if $P(B/A) = 1$, it does not follow that $P(\sim A/\sim B) = 1$, since if $P(B) = 1$, $P(\sim A/\sim B)$ is undefined. Nor does contraposition hold in standard conditional logics. Therefore, if the connective of a law is not a material conditional, but a strong conditional of some kind, as I would suppose, then, quite apart from any considerations about natural kinds, we should be able to distinguish between laws and their contra-positives.

Let us therefore distinguish between the positive and contrapositive formulations of laws, and define the *scope* of a law as the set of all things in the class to which reference is made by the antecedent term in its positive formulation. This concept of scope may then be used as the basis for an ontologically revealing way of classifying laws. For the scope of a law defines the set of things it is about, and hence the set of things on whose properties the law must depend. We shall say that the law *applies* to these things.

Given this concept of scope, we may now usefully distinguish between laws on the bases of (a) their generality and (b) the kinds of things to which they apply. First, the most general laws are the *global laws,* which apply to all events and processes. These are exemplified by the conservation laws. Every event or process that can occur is *intrinsically* conservative in a number of respects. Of course, most events and processes occur in *open* systems – systems that are subject to the action of forces, or to or from which there is some flow of matter or energy. But when corrections are made to allow for such extrinsic factors, it can be seen that the conservation laws apply unrestrictedly to all events and processes. Every event or process that can occur in nature is intrinsically conservative of energy, momentum, angular momentum, charge, hypercharge, and so on.

Second, there are *general structural principles,* which apply to structures. These laws define the space-time or energy structure of worlds like ours, and what kinds of states must or cannot coexist in such a structure. Like the conservation laws, these principles are global. They do not refer to any particular kinds of things or substances, but apply universally to all kinds of things or substances. The principles of relativity, for example, do not discriminate between different kinds of objects. Planck's law states a general equivalence between the energy of a thing and the frequency of the associated wave. Heisenberg's Uncertainty Principle and Pauli's Exclusion Principle impose certain absolute restrictions on the coexistence of states.

Third, there are laws that are concerned with the essential natures of more specific kinds of objects or substances, such as particles, chemical substances, and fields. That electrons are negatively charged, for example, is a law of this kind. So too are the chemical facts concerning the structure of copper or silver compounds. It may be doubted whether such propositions are really *laws*. David Armstrong, for one, does not think they are.[3] But the laws of electromagnetism that are expressed by Maxwell's equations would also appear to be propositions of this kind. For these propositions do nothing more than describe the essential nature of the electromagnetic field. They are not laws that are global in scope like the general structural principles, but they are surely laws of nature. Plausibly, propositions describing the essential properties of other kinds of fields, such as those dealt with in quantum field theory, also belong to this category.

Finally, there are the causal and statistical laws that tell us how things of various kinds are disposed to behave or interact with each other. These laws are more limited in scope, and apply to specific kinds of events or processes. They do not apply globally to all kinds of processes, as the conservation laws do, but are restricted in their application to a range of specific kinds of things. The laws of chemical interaction, for example, apply to the specific chemicals involved. They are not global in scope[4] in the way the conservation laws are.

6.3 THE CAUSAL LAWS

The causal laws[5] apply to the natural kinds of processes that can occur in the world, and are grounded in the causal powers of the things involved in them. A causal power, we have argued, is a property that an object may have in virtue of a relationship between that object and a natural kind of process. That is, it depends ontologically on a relationship between the object and a generic dynamic universal. The natural kinds of processes involved in such relationships are the displays of the causal powers involved. These processes begin with events of the natural kinds that activate the causal powers (the causal kinds) to produce correlative events of the natural kinds that result from such activations (the effect kinds). The essential properties of the fundamental natural kinds in the category of substances include a number of different kinds of causal powers. Things of these kinds must therefore be intrinsically disposed to act as these powers require, and must so act if the activation conditions are satisfied. The laws that describe how things with various causal powers are intrinsically disposed to act in virtue of having these powers are the causal laws of nature.

If this is right, then it follows that the causal (including statistical) laws concerning the behavior and interactions of things belonging to natural kinds are not mere contingencies. For the things belonging to these kinds must have the intrinsic dispositional properties they have to be the things of the kinds they are, and to have these dispositional properties, they must be disposed to behave and interact as they do. For this is just what it is to have these properties. Therefore it is not a contingent matter that electrons repel each other or that hydrogen and chlorine combine to form hydrogen chloride. There is no possible world in which they would not do so. If the laws of electricity were different, then the world would not contain electrons. If the world contained substances like hydrogen and chlorine, but they could not be combined to form hydrogen chloride, then these substances would not be hydrogen and chlorine. In defending this view, I will be taking sides in an old dispute about the sources of power in the world. For the view that the various kinds of substances that exist have intrinsic causal powers, such as I am supposing hydrogen and chlorine to have, is a view that was attacked and vehemently rejected by the British empiricist philosophers of the eighteenth century. These philosophers thought nature to be fundamentally inert or passive and therefore incapable of acting. This conception of reality, and the difficulties associated with it, will be discussed in the next chapter. It is, I shall argue, a metaphysic that has blocked all serious attempts to develop a satisfactory theory of laws.

For the remainder of this chapter, I will be concerned with the problems one would expect any adequate theory of laws of nature to be able to deal with. I shall argue that other accounts of laws fail on one or more counts to deal satisfactorily with these problems.

6.4 DESIDERATA FOR A THEORY OF LAWS

Most philosophers would agree that laws are not just de facto or accidental regularities. How then are laws to be distinguished from such regularities? The problem is usually posed as though it were a formal one of distinguishing between two kinds of universal statements – *nomic* and *accidental?* In particular, it is asked: What are the distinguishing features of those universal generalizations that entitle them to be called laws of nature? The presupposition of posing the question in this way is that laws are universal generalizations of some kind. The problem is just to specify the kind. I am unhappy with this way of putting the question, however, because it suggests a much closer link between accidental and nomic universals than I am prepared to admit. Of course, I agree that law-statements are universally quantified propositions. But I do not agree that they can adequately be represented as universal generalizations in first-order predicate logic, as accidental generalizations are. Law-statements, I shall argue, are fundamentally modal in a way that accidental generalizations are not.

Philosophers of science have made many attempts to specify necessary and sufficient conditions for lawhood. As might be expected, none of the criteria that have been proposed is entirely beyond dispute. Nevertheless, some criteria are very widely accepted. Statements of laws must (in some sense) be universal, explanatory, predictive, and bear some relation to science. Most of these have been discussed at length elsewhere,[6] and, although not universally embraced by all accounts, are largely uncontroversial. There are, however, three criteria that deserve special mention, because of the problems they create. There is also a problem of demarcation (not Popper's problem, but another one) between genuine laws of nature and formal principles of the kind that are to be found in formal sciences, such as Euclidean geometry and neoclassical economics.

Arguably, the most important, and certainly the most puzzling, characteristic of laws is their apparently special brand of necessity. It is almost universally agreed that laws of nature are necessary in some sense. However, laws are also *a posteriori*, and so, presumably, they are not logically necessary. It is supposed, therefore, that laws of nature must be necessary in some weaker sense of "necessity." But what could this be? Presumably, the laws of nature are physically necessary. But then how is physical necessity to be defined? Manifestly, it will not do to say that the physically necessary propositions are just those that are true in all worlds governed by the same laws as ours. For this would be to argue in a circle. One might as well try to define a historically necessary proposition as any proposition that is true in all worlds with the same history as ours.

The first desideratum for a theory of laws, then, is that it should give an adequate account of natural necessity. I call this the *Necessity Problem*. Van Fraassen draws attention to the difficulty of giving an adequate account of natural necessity. As he construes the problem, it is that of identifying the relationship in virtue of which an empirical generalization acquires nomic status. Accordingly, he calls this difficulty "the problem of identification." However, in so construing the problem, he begs an important question. For, his way of posing the problem presupposes that natural necessitation is a relation that somehow imparts its character to what would otherwise be just an ordinary empirical generalization. I reject this presupposition. Natural necessity is not like greatness. Laws of nature do not achieve necessity, or have necessity thrust upon them; they just are that way.[7]

One common and important property of laws that has received very little attention in the philosophical literature is their abstract or idealized nature. There are very few laws that apply directly to the kinds of things or processes that are actually observable in the world, and those that do are generally regarded as low level empirical generalizations. Most of the propositions we think of as being (or as expressing) genuine laws of nature seem to describe only the behavior of ideal kinds of things, or of things in ideal circumstances. This common feature of laws has often been noted (for example, in Cartwright, 1983[8] and Ellis, 1992a), but most theories of laws do little or nothing to explain it. Many writers have supposed that the fact that idealizations occur in so many laws reflects only our need to simplify nature in order to understand it.[9] Nature is too complex, it is said, for us to be able to formulate basic laws that apply directly to it. Our laws are therefore a kind of compromise between truth and intelligibility. As statements about reality, they are at best only approximations to the truth.

I think that this account of idealization in science is unsatisfactory since it does not explain why highly idealized theoretical models should often be preferred. Moreover, they are evidently preferred even when more realistic models are available. (Think of black body radiators and perfectly reversible heat engines.) This is what I call the *Idealization Problem*. In meteorological and economic forecasting, accuracy of prediction is important, if not always achievable. Consequently, to the extent that economic and meteorological models do not accurately reflect reality, they are unsatisfactory. In these fields, therefore, our theoretical models are expected to be as realistic as possible. But weather and economic forecasting are not typical sciences. Typically, the emphasis in science is not on forecasting, but on understanding. And for reasons that are not well understood, this often seems to require high levels of abstraction and idealization and the construction of models that are known to be unrealistic.

Another property any decent theory of laws should be able to account for is objectivity. That is, the laws of nature should be discoverable, and therefore be or describe some reality which exists independently of us. This is so, I think, even if it can be argued that there are conventional elements in some of the laws we accept. For the laws of nature are clearly not just inventions or abstract mathematical constructions. Many of them, at least, are postulates about the structure of reality or the kinds of processes that can occur in nature. This being so, we should be able to say precisely what features of reality the laws of nature describe. This would be straightforward enough if a Humean theory of laws were defensible and the necessity of laws could be adequately accounted for independently of their descriptive roles. For then their basis in reality would be just the universal regularities they describe. However, a Humean theory of laws is not defensible, and the usual accounts of natural necessity are unsatisfactory. But if a Humean regularity theory is rejected, then there is a problem concerning the ontological foundations of laws – namely, in what features of reality are the laws of nature grounded? I call this the *Ontological Problem*.

It seems, then, that there are at least three major problems about laws of nature – the *Necessity Problem,* the *Idealization Problem,* and the *Ontological Problem.* Any adequate theory of laws should yield satisfying solutions to these. Currently accepted theories of laws do not do so – or so I shall argue. Humean regularity theories do not provide acceptable solutions to the Necessity or Idealization Problems; they are unable to explain the necessity of laws, nor can they account for the idealizations that so often occur in laws of nature. However, such theories are compatible with a Humean ontology, and for this reason, philosophers who accept a Humean ontology may consider it to be the only kind of theory that solves the Ontological Problem.

Conventionalist theories of laws fare better with the Necessity and Idealization Problems, but fail to provide a satisfying ontology of laws. They do not explain the objectivity of laws of nature, and they do not provide truthmakers for them. The natural necessitation theories of F. I. Dretske (1977), M. Tooley (1977), and D. M. Armstrong (1983) may seem promising for dealing with the Ontological Problem. But they fail on the Necessity and Idealization Problems. C. Swoyer's (1982) theory of laws of nature is probably the best in the field. It gives a plausible account, similar to that offered by Dretske, Tooley, and Armstrong, of the truthmakers for laws, and it also explains their necessity. However, Swoyer's theory, by his own admission, fails to deal adequately with the Idealization Problem.

Like Swoyer, I think that the only kind of theory of laws of nature that can solve both the Necessity and Ontological Problems is an essentialist one – a theory on which the basic laws of nature are really (metaphysically) necessary. But, I argue, the Idealization Problem can only be solved as well if we have a richer ontology than Swoyer's – indeed, one that is richer than that of any of the other natural necessitation theorists. The ontology accepted by these philosophers is one of universals (properties and relations) and particulars (the bearers of these properties and relations). What is needed, I think, is an ontology, like the one described in Chapter 2, which also includes some fundamental, irreducible, categories of natural kinds. In this chapter, I shall discuss the three main kinds of theories of laws that are rivals to the theory being developed here, and explain how these other theories of laws come to grief by their failure to solve the three basic problems about laws of nature.

Hume's theory about the laws of nature has come to be known as the *Regularity Theory.*[10] According to Hume, the laws of nature are causal laws, and causal laws are just regularities of some kind. They are not necessary, except in the sense that they are felt by us to be so. From the appearance of the cause, we may come, by habit-forming experience, to anticipate the effect. So the effect naturally seems to us to be produced by the cause, and hence necessitated by it. But the necessitation is really just an illusion created by our anticipation of the effect. It does not exist in reality.

Many philosophers who are otherwise sympathetic to Hume's philosophy have not been happy with Hume's account of natural necessity – the sort of necessity that is supposed to characterize causal laws. To say that one kind of event, say A, is the cause of another, say B, is not just to say that whenever an A occurs a B will occur. For this might be true just by chance. For one thing to be the cause of another, it is argued, it must in some sense bring it about; so that if circumstances of the kind A were to recur, an event of the kind B would then have to occur. But a Humean, it seems, cannot have the required concept of causal power because, from a Humean perspective, there is no such thing as a necessary connection between events that is conceivable by us.[11]

The same is true of all laws of nature. If it is a law that all As are Bs, then it is not merely true that all As happen to be Bs. Any further As that might be produced would also have to be Bs. For example, there are in fact no perpetual motion machines of the first or second kinds. But the First and Second Laws of Thermodynamics, which forbid them, tell us more about the world than this. They tell us that there could not any such machines. It is physically impossible for there to be, or for anyone to construct, such machines.

Humean theories of laws also have trouble with the Idealization Problem. What can a Humean say about the law L, which states that the efficiency of any perfectly reversible heat engine working between temperature limits t_1 and t_2 is the same, and is greater than that of any irreversible heat engine working between these same temperature limits? Considered as a universal generalization, it is vacuously true. In practice, it is not possible to build a heat engine that comes even close to being perfectly reversible, and the efficiencies of real heat engines are a long way below their theoretical maxima. A Humean might try the suggestion that L is not really a law but a definition. However, it does not appear to be a definition of perfect reversibility, or of efficiency, or of any of the other terms used in its statement. Perhaps, then, it is like a proposition of Euclidean geometry. Maybe. But then a Humean is obliged to deny that L is a law of nature. For the laws of nature are supposed to be regularities of some kind, not theorems of abstract theoretical systems.

One common Humean strategy for dealing with the Idealization Problem is to argue that ideal laws such as L are really compromises between the competing demands of accuracy and comprehensibility (Scriven, 1961). The true laws, which apply directly to reality, would often be far too complicated to be stated in an intelligible way. Therefore, we must make simplifying assumptions, and make do with approximations that we can grasp and work with. I call this "the approximation defense." However, there are good reasons for thinking that this is not the motivation for idealization in science. First, ideal laws often remain the fundamental ones, even when much more realistic laws are known. The perfect gas laws, for example, are still the fundamental laws of the theory of gases, even though real gases are not perfect, and are known to behave in other ways – more or less as Van der Waals' equation of state implies. However, the theory of perfect gases remains the basic theory, and Van der Waals' equation of state is just a modification of it that is of no great theoretical interest. It is not that Van der Waals' equation is very complex. On the contrary, it is quite simple. Van der Waals' equation is not discussed very much in physics textbooks, simply because it is not very interesting.

Second, even some of the most fundamental laws of nature are abstract. The conservation laws, for example, all refer to idealized systems. These laws are supposed to hold exactly only for systems that are closed and isolated. They are not thought to hold precisely for the kinds of open and interacting systems we find in nature. Therefore, a Humean who adopts the approximation defense is obliged to say that these laws too are a compromise between the demands of accuracy and comprehensibility. This is surely nonsense. These are the fundamental laws of nature, not approximations to the true laws adopted as a kind of compromise.

Of course, the universe itself is a closed and isolated system. So it cannot be said that the conservation laws are all vacuously true. But the conservation laws also tell us a great deal about what happens locally, and would be useless if they did not do so. Therefore, any satisfactory account of the laws of nature must deal adequately with the Idealization Problem, which is raised in a very acute form by the conservation laws. The fact that the conservation laws all apply globally is not much of a consolation to a Humean. The regularity theorist must be able to explain how the conservation laws can apply locally, even though none of the open and inter-active systems we find in nature actually obeys them.[12]

Another possible line of defense for a Humean would be to argue that the idealized laws we find in science express deep regularities – regulari-ties that may never, or very rarely, appear at the surface level. But a Humean who takes this view immediately runs into difficulties with the Ontologi-cal Problem. For what are the idealized objects involved in these deep reg-ularities? Are there really, deep down, closed and isolated systems, inertial frames, perfectly reversible heat engines, and the like? If not, then how can these deep regularities exist?

6.7 CONVENTIONALIST THEORIES OF LAWS

At the turn of the twentieth century, Henri Poincaré and a number of philosophers, most notably Ernst Mach, defended the view that many of the most fundamental laws of nature are not just empirical generalizations, as Hume had supposed, but conventions adopted because of their con-venience in organizing and systematizing experience. The experimentally discovered laws that our theories are designed to explain were of course held to be empirical generalizations. But many of the basic laws of nature that are embedded in our explanatory theories were held to have a dif-ferent status. These laws cannot be verified or refuted experimentally, it was held, for the simple reason that they are not experimental laws. They are conventions adopted because of their utility as components of theo-ries. Consequently, they would cease to be accepted only if they ceased to be useful for this purpose, or if more useful conventions could be pro-posed to take their place. Laws of nature having the status of conventions might stand or fall with the theoretical structure that they support, but not independently of it.

According to Mach and Poincaré, the laws of nature have a status similar to that of a geometrical theorem. They would say, for example, that the law of inertia is true in Newtonian mechanics (since it is an ax-iom of that system), just as they would say that Playfair's axiom is true in Euclidean geometry. They would not say of either axiom that it is true absolutely since other geometries and other systems of mechanics can be developed to describe the same reality. The laws of mechanics, like the laws of geometry, are conventional.

This conventionalist theory of laws gains plausibility and support from the fact that many laws, especially many of the most fundamental laws, apply strictly only to ideal systems existing in various kinds of ideal circumstances. For such laws can easily be construed as conventions. Conventionalists would say, for example, that the law of conservation of energy is a convention – one that serves (at least partly) to define the concept of a closed and isolated system. Therefore, they would say, there cannot be any exceptions to the law of conservation of energy. If we come across a system for which energy is not conserved, then this only shows that it is either not closed or not isolated.

As a theory of laws of nature, conventionalism has some good things to be said for it. First, it accounts well for the *a posteriori* necessity of basic laws. These laws are necessary, because they are true by definition or convention – that is, they are necessary *de dicto*. On the other hand, since we have to discover by trial and error what conventions are best for our theoretical purposes, it is *a posteriori* what conventions we should adopt. Moreover, the conventions we adopt must be chosen specifically for the purposes of theory construction, and can continue to be accepted only if they continue to be useful for this purpose. If the theory is superseded by a better theory in which other, more useful, conventions are adopted, then the laws are considered to have been falsified (although, strictly speaking, they have just become obsolete).

Second, conventionalism accounts well for the abstractness and ideality of laws. If the laws of nature have a status similar to geometrical theorems, then they will be concerned with abstract, idealized entities similar to those of geometry. They cannot be concerned directly with real objects or processes, for then they would be open to empirical refutation. If the axioms or theorems of a physical theory were empirically testable, then they would not have the status of axioms. What is testable is only whether the theory can be applied as it was intended to be.

There are, however, serious problems with conventionalism. First, some basic laws of nature do not appear to be just conventions. If one focuses on Newton's laws of motion, or on the principle that the speed of light is the same in all directions, then a fairly good case for conventionalism can be made out. Alternative, even if somewhat less convenient, theoretical frameworks can be constructed to do the same work.[13] The law of conservation of charge, on the other hand, which is surely a fundamental law, cannot plausibly be construed as a convention, and it is hard to see how it could reasonably be defended against direct counterevidence.

Second, conventionalists can offer no satisfactory account of why some "conventions" are so successful and so much better than any others. If the laws of nature are just conventions, how can we explain the fact that (with very few exceptions) there are often no viable alternatives to the conventions we have adopted? We can argue abstractly – for example, from the empirical underdetermination thesis – that logically distinct but empirically equivalent theories based on different conventions are always possible. However, genuine examples of such alternatives are hard to find (Ellis, 1985). It is easy enough to construct an alternative theory T'to a theory T, if T' = Even though not T, the world behaves (appears to us, etc.) as if T. However, T' is not a genuine alternative to T.

Third, if the necessity of laws is just *de dicto* necessity, then where is their basis in reality? *De dicto* necessities derive from the conventions of language, not from the ways of the world. Surely any satisfactory account of natural necessity would ground it in nature, not in how we choose to talk about it. The conventionalist theory of laws thus fails to provide a satisfactory resolution to the Ontological Problem.

6.8 NATURAL NECESSITATION THEORIES

As a result of the failure of Humean and conventionalist theories of laws to deal adequately with the problem of natural necessity, many philosophers have sought a more realistic basis for this kind of necessity. On any satisfactory theory, it is argued, the laws of nature must turn out to be necessarily true in virtue of some real relation of natural necessitation. Let us call any such theory a *natural necessitation theory* of laws. The theory of laws I wish to defend in this book is one such theory. However, the theory to be defended differs from some other natural necessitation theories in at least one important way. Like Swoyer (1982), I deny that the laws of nature are contingent. But many of those who have argued for natural necessitation theories have tried to have their cake and eat it too. On the one hand, they want to say that the laws of nature are necessary, while on the other, that they are contingent. I call any such theory of laws a "contingent natural necessitation theory." Those who adopt it are forced to defend both the contingency and the necessity of laws, and consequently they must, in the end, try to defend the view that natural necessity is a kind of hybrid between contingency and necessity. The laws of nature, they are forced to conclude, must in some sense or at some level be contingent, while in another sense or at another level they are necessary. John Bigelow, Caroline Lierse, and I do not think that this is a viable position, as we argued in Bigelow, Ellis, and Lierse (1992).

Perhaps the best known theories of contingent natural necessitation are those from F.I. Dretske (1977), M. Tooley (1977), D.M. Armstrong (1983), and J.W. Carroll (1984). According to these theories, the laws of nature are

215

relations of natural necessitation between universals. If it is a law that all As are Bs, then this must be because there is relation of natural necessitation between two genuine universals – A-hood and B-hood. Given that these universals exist, and that this relation holds, then it follows necessarily that if any instance of the former occurs, then an instance of the latter must also occur.

Contingent natural necessitation theories also have some good things to be said for them. For example, they make natural necessity a species of *de re* necessity. But the kind of theory outlined will not do as it stands. As we have argued elsewhere,[14] the natural necessitation relation that is supposed to link the universals is mysterious. On Armstrong's formulation, this law-making relation is an irreducible, dyadic, contingent relation of natural necessitation (or probabilification) holding between the universals concerned. This gives Armstrong what he wants for the kind of case he is considering – contingent laws that entail the corresponding universal generalizations but nevertheless have the kind of necessity he is seeking to explicate. But the necessity is no less mysterious for being considered to be characteristic of the relation between the universals. What makes this relationship necessary? It is not a relation between the universals themselves, which plausibly would be a necessary relationship (since universals are not world-bound). For this would make the laws true in all possible worlds. No, it has to be a relation that just happens to hold in our world. And this is very mysterious.

The main trouble with all of these theories is that they seek to combine what is basically an Aristotelian theory of natural necessity with a Humean contingency thesis about laws. It cannot be done. There is no possible account of natural necessity that is compatible with the view that the laws of nature, and hence the law-making relation, is contingent. The problem of reconciling real relations of natural necessitation with a Humean metaphysic has not been, and cannot be, solved. If the relation between the universals is contingent, then the laws are contingent; if it is logically or metaphysically necessary, then the laws are logically or metaphysically necessary. If the relation of natural necessitation is some other kind of necessitation relation, then we are owed an account of it. It seems to me that no analysis of laws can reconcile the claim that the laws of nature are contingent with the view that they are, in some robust sense, also necessary. No amount of metaphysical maneuvering will escape this difficulty. What is needed is something more radical – a thoroughly non-Humean theory of natural necessitation.

A non-Humean theory of natural necessitation would ground this relation in the dispositional properties of things – that is, in their causal powers, capacities, and propensities. The essentialist theory of laws, which is to be developed, is a natural necessitation theory of this kind. This theory will be fully explained in the next two chapters, where it will be argued that most general laws of nature – the global laws – describe the spatio-temporal and causal structure of worlds of the kind we live in, while the more specific laws, including the causal and stochastic laws, describe the natural kinds of processes that can occur in worlds like ours.

It will also be argued that the natural kinds of processes that can occur are generally the displays of the basic dispositions of things, their causal powers, and so on. And to the extent that this is the case, the laws of nature must be grounded directly in these properties rather than in any higher order relations of natural necessitation holding contingently between properties. So at least in these cases, my theory is like Swoyer's (1982), and is similar to the theories proposed earlier by Harré and Madden (1973) and Shoemaker (1980). For, according to all of these theories, the laws applying to things of these kinds are directly grounded in their natural dispositional properties. They are grounded in these properties in the sense that they (the properties) are their truth-makers.[15] Whatever has these properties must be disposed to behave according to these laws, because this is what it is to have these properties.[15]

The properties that are the truth-makers for the causal laws are the causal powers of things.[17] Accordingly, anything that has a causal power essentially must be disposed to behave as it does when this power is activated, not because of anything extrinsic to it (such as the laws of nature, or relations of natural necessitation between universals), but because, in virtue of its being a thing of this kind, it is intrinsically disposed to behave in this way.

According to scientific essentialism, at least some dispositional properties are fundamental – not ontologically dependent on categorical (non-dispositional) properties or on the laws of nature. These fundamental dispositional properties of things are the truth-makers for the most fundamental causal laws that ultimately determine the ways in which things are disposed to behave, or with what probabilities they will be so disposed. The displays of these various dispositional properties are the instances of the most fundamental natural kinds of processes. We do not say that there are no fundamental categorical properties. On the contrary, we think that spatio-temporal and numerical relations, at least, are categorical and also fundamental. We do not see any benefit in trying to reduce these relations to dispositional properties. Nor do we think it is possible to do so. Such categorical relations as these would appear to be essentially involved in every natural kind of process.

There are, I think, two broad categories of fundamental natural processes — *basic causal interactions* and *energy transmission processes*. The basic causal interactions are the composite displays of the dispositions of the various kinds of objects involved in them. They include the interactions between particles and between particles and fields, including the making of measurements, and generally all of those processes that involve collapsing wave packets. These interactions may well be non-local, as Bell's theorem demonstrates.[18] The energy transmission processes, on the other hand, include all of those processes by which energy is transmitted from one place to another — for example, inertial motion and electromagnetic radiation. An essential difference between the two kinds of processes is that the energy transmission processes are limited by the velocity of light, and hence cannot have non-local effects, as causal interactions may.

The natural kinds of processes that causal laws describe are not like natural kinds of objects in at least one important respect — they can occur on top of one another. Two objects cannot occupy the same place at the same time. But a single object can undergo several processes simultaneously. Consequently, there may be natural kinds of processes that seldom occur naturally in isolation from other processes, and are often obscured by them. Indeed, some natural kinds of processes may never occur in isolation from other processes. Therefore, to describe a natural kind of process, it may be necessary to abstract from the real-life situations in which such a process may actually occur, to consider what would happen in those highly contrived or idealized circumstances in which the process would occur without interference. Consequently, we should ordinarily expect laws describing natural kinds of processes to refer to ideal objects and circumstances. For these laws must tell us how the natural dispositions of these objects would be seen to be displayed, if they were not subject to any perturbing influences. We should also expect experimental design to focus on ways of isolating the phenomena to be investigated and eliminating extraneous forces. For if the laws we are seeking are descriptions of natural kinds of processes, then we need to study them in their purest form, or as near as we can get to such a form.

Given that we think the world itself is one of a kind, it follows that we should think that it has certain global properties and structures that are responsible for its being the kind of world it is. For example, we should suppose that there are certain kinds of ways in which all things in this world, and in all worlds of the same natural kind as ours, are intrinsically disposed to behave. Thus, I conjecture that all things are intrinsically disposed to behave in ways that are conserve energy, momentum, angular momentum, charge, and so on. But of course all things, or at least all that we can know about, are interactive in some ways, and many are constantly interacting in many ways. Consequently, the laws that express these behavioral dispositions must be idealized, as indeed all of the conservation laws are. That is, they must refer to how the various things we have to deal with would behave if they were causally isolated, not to how they really do behave in the causally open systems in which they actually exist.

6.10 SOLVING THE THREE MAIN PROBLEMS

By construing causal laws as descriptions of natural kinds of processes, an essentialist theory can offer satisfactory solutions to the three main problems.

The Necessity Problem

An Essentialist theory can offer an adequate account of the necessity of laws. For it is a necessary truth that a thing of kind K has the property P if P is an essential property of K. It is, of course, *a posteriori* what properties are essential to a given kind. Therefore, the proposition that things of the kind K have the property P is what I call "really necessary." If P is a natural dispositional property, then it is also a necessary truth that anything having the property P must be disposed to behave in certain ways in certain circumstances just in virtue of having this property. Of course, we have to discover empirically what kinds of dispositional properties exist. But if anything has the property P, it must be disposed to behave in a P-wise fashion. And if anything of kind K has the property P necessarily, because P is an essential property of K, then it must be disposed to behave in a P-wise fashion, just in virtue of being a thing of this kind. Therefore, if the laws of nature are propositions stating facts of this sort, then they too are really necessary.

According to scientific essentialism, the laws of nature have this character and metaphysical status. Their epistemic status is of course *a posteriori*. So the laws of nature we are concerned with here are both really nec-

219

essary and *a posteriori*. They are thus a bit like logical truths and a bit like empirical truths. They are necessary, as logical truths are, but like empirical truths, they have to be discovered by the methods of empirical science.

To many philosophers, this conclusion will be radically unacceptable. If the causal laws we are discussing are really necessary, then it is impossible, metaphysically impossible, for anything to behave in ways other than in those required by these laws.[19] The only degrees of freedom in what might happen are just those allowed by quantum uncertainties, and these degrees of freedom, it will be said, are simply not enough to explain our common intuition that things contrary to the causal laws of nature logically could happen. Therefore, it is concluded, natural necessity must be weaker than real necessity. It cannot be a contradiction to suppose that anything contrary to the causal laws of nature should occur. That is just too strong. The contrary of such a law will certainly be false, and perhaps, in some weaker sense than mine, necessarily false. But it will not be impossible in the sense that it cannot occur in any possible world. It is true that most suppositions about what might happen are made in ignorance of the exact constitutions of the things we are dealing with and without knowledge of the causal laws concerning their constituent parts. Consequently, it is often easy to believe that any of a number of things might happen to these things in the existing circumstances. They are all epistemically possible. So it is tempting to argue that if the alternatives to what actually happens are all epistemically possible, then their actually happening must be at least logically possible, and therefore that the causal laws concerning the behavior of these things cannot be really necessary.

To this objection I make the following replies (a) Epistemic possibility does not entail logical possibility. It is epistemically possible that the number that consists of the first million digits in the decimal expansion of π is the product of two consecutive primes. For all anyone knows, this is true. But if it is false, then it is necessarily false. Therefore, what is epistemically possible may be logically impossible. (b) While it is a necessary truth that things of the natural kind K will be disposed to behave in a P-wise fashion, if P is an essential property of K, it is not epistemically necessary that a given thing, say *a*, is an instance of K. It is true that if *a* is an instance of a natural kind K, then *a* must behave in this fashion.[20] If *a* is not a silver atom, although we mistakenly think it is, then our belief that this very atom is a silver atom is necessarily false. It may not of course be irrational for us to believe this. For, on the evidence available, it may be quite reasonable to believe that the thing in question is a silver atom. But this does not mean that it is contingent whether the thing I am actually observing is or is not a silver atom. What is not a silver atom could not be (or become) a silver atom, whatever we might think or suppose.[21] We can be mistaken about what kinds of things there are, and also about what belongs to them. But these facts imply only that it is epistemically contingent whether what we are actually observing is or is not a thing of the kind we think it is. It does not imply that it is really contingent what sort of thing it is.

Why then are we all so easily misled into thinking that what is really impossible is nevertheless possible? I think it is because our judgments of possibility are normally based either on what we are able to imagine – cartoon fashion – happening, or else, just on ignorance. For if we cannot think of a good reason why something cannot happen, or be the case, then we are likely to infer that it is at least logically possible for it to happen, or be the case. But to base judgments of possibility on such considerations is to suppose that epistemic and representative possibilities imply real possibility, which they do not. Judgments of real necessity and possibility should not be based on ignorance or imagination. Rather, they should depend on our being able to identify the kinds of things we are concerned with, and discover what their essential properties are. For once we have this knowledge, then we could, at least in principle, find out for sure what is really possible and what is not. This seems to me to be the better way of determining what is possible, even if the judgments that we should make based on a knowledge of the natural kinds involved, and their essential properties and structures, would often be at variance with those based on our powers of imagination, or on our ignorance of real possibilities.

The Idealization Problem

The essentialist theory of causal laws explains the ideality of many such laws, and so resolves the Idealization Problem. For the kinds of processes with which we are concerned may never, or rarely, occur in causal isolation. Consequently, to formulate these laws, it may be necessary to abstract from the kinds of circumstances that may actually prevail in order to consider what would happen in the idealized circumstances in which the processes in question would occur without interference. This is a simple consequence of the fact that many different kinds of processes can, and often do, occur to the same things at the same time. This point will be discussed more fully in the next section.

221

Scientific essentialism offers a satisfactory ontology of laws. For, according to the essentialist theory, the causal laws of nature are objective, and describe the essential natures of the basic kinds of causal processes occurring in nature. In all cases, the causal laws derive from the causal powers, capacities, or propensities of the participants in these causal processes. These powers, and so on, are therefore the truth-makers for these laws. This point will be discussed further in the next chapter. Scientific essentialism thus explains the necessity of laws as a species of *de re* necessity, and so not the kind of necessity that is dependent on the conventions of language. It solves the Idealization Problem by relating the search for causal laws to the quest to discover the real essences of the natural kinds of processes that can occur in worlds like ours. And it solves the Ontological Problem by positing plausible truth-makers for laws of nature, thus explaining their objectivity.

Finally, the essentialist theory of laws can be shown to have a unique advantage over all other theories: It explains their hierarchical structure. For it grounds the hierarchies of laws in the hierarchies of natural kinds of objects and processes. An ontology that does not have such natural hierarchies cannot do this. At best, it can only be argued that since there are natural hierarchies of laws, there must be hierarchies of kinds of objects and processes. But this is to put the ontological cart before the horse. It is true that to derive the conclusion "Necessarily, Ss are Ps" from the premise "Necessarily, Gs are Ps" we need to know that necessarily Ss are Gs, so that if S and G are natural kinds, what we need to know is that S is a species of G. But these species relationships must exist if these derivations are to be sound. Therefore, to explain the existence hierarchies of laws, we need an ontology in which there are hierarchies of natural kinds of objects and processes to which they apply.

6.11 THE ABSTRACT CHARACTER OF THEORETICAL LAWS

It is often necessary for science, in pursuit of its aims, to abstract from the accidental properties of things, and the extrinsic forces that act on them, to consider how they would behave independently of these properties, or in the absence of these forces. For this is the only way of finding out what behavior is generated by the intrinsic properties and structures of the kinds of things we are studying, and what is due to other, extraneous, influences. And this interest would remain the primary one, even if we could determine precisely the effects of the accidental properties of things, in the circumstances in which they actually occur. For the aim of science is not to describe what actually happens in nature, or to systematize our knowledge of what occurs by subsuming it under general laws that are seen manifestly to be obeyed; rather, it is to explain what happens by showing how it can be seen to arise out of the essential natures of the natural kinds of things that constitute the real world.

We have seen that natural kinds are to be distinguished from each other by their intrinsic properties and structures.[22] Copper and gold, for example, are to be distinguished from each other, and from all other substantive natural kinds, by their distinctive nuclear and electron structures. The causal powers that necessarily belong to the individuals that instantiate the natural kinds are ultimately to be explained by reference to these properties and structures. Such an explanation is what I have called an *essentialist explanation,* because the intrinsic properties and structures required for this explanation are just the essential properties of the natural kinds involved.[23]

It has been argued at some length (in Chapter 4, especially in Section 4.5) that there are natural kinds of processes, as well as natural kinds of objects. By natural kinds of processes I mean ontologically objective kinds of events or sequences of events, having certain intrinsic properties or structures that make them the kinds of events they are, and distinguish them categorically from all other kinds of events or sequences of events. It was also argued that natural kinds of processes are among the primary objects of scientific inquiry. In fact, it is plausible to suppose that the principal aim of natural science is to discover what natural kinds of processes can occur in this world, and what their essential natures are.[24]

Inquiry in the physical sciences is largely concerned with a species of natural kinds of processes – namely, causal processes, that is, causal interactions connected by energy transfer processes. The kinds of causal processes of greatest interest are, as might be expected, those involving the natural kinds of things that are the chief objects of physical scientific inquiry. The interest lies in how and why things of these kinds are disposed, as they are, to behave or interact with each other, or with things of other kinds. Perhaps the kinds of causal processes that are of greatest interest in the physical sciences are those that are driven by the intrinsic natures of their participants. It may be that such processes do not often occur, or do not occur except in certain circumstances (for example, two substances may have to be mixed in solution for a reaction to take place), or that they rarely or never occur in the absence of external forces (which may modify or distort the internally driven sequences of events which characterize these processes). Nevertheless, it is clear that there are causal processes that are driven by the intrinsic natures of the things directly involved in them, and that consequently these processes have a certain intrinsic nature. These will all be instances of natural kinds of causal processes.

6.12 THE ROLE OF IDEALIZATION IN PHYSICAL THEORY

To suppose that idealizations occur in physical theory just because reality is too complex to be dealt with otherwise is to misunderstand their role. For even where we are able to develop more realistic models of natural processes than the simple models currently accepted as adequate for scientific purposes, physical theorists are often not much interested in them.[25] The reason is that the aim of physical theory is not just to describe nature, or to develop theories from which true universal generalizations about how things behave in nature can be derived. Rather it is to discover what kind of world we live in, what kinds of things exist, and what their essential properties and causal powers are.

The kinds of theories and explanations sought by physical theorists are directly related to these aims. The most general physical theories, such as our theories about space-time, and the theories we have about the interrelatedness of the fundamental forces (unified field theories), are concerned with the kind of world in which we live. These theories are attempts to describe the spatio-temporal and energy structures of such a world, and if they are true, then they are necessarily true of worlds like ours. A world could not lack these structures, yet be a world of our kind.

The elementary causal processes that can and do occur in worlds like ours include decay processes, causal interactions, and energy transmissions. These three kinds of processes are intrinsically conservative of various quantities — those for which there are conservation laws. Since these laws are concerned with the intrinsic nature of the elementary processes of nature, they can apply strictly only to causally isolated processes — processes that are not influenced by any external forces. Therefore, to state these laws properly, it is necessary to refer to idealized processes that may be supposed to occur in isolation from any outside causal influences. Thus, the laws of conservation of energy, momentum, angular momentum, charge, lepton number, baryon number, and so on all apply strictly only to closed and isolated systems. But apart from the universe itself, real systems are never closed and isolated. They are all, for example, acted on by gravitational and electromagnetic forces. Therefore, to state some of the most fundamental

224

laws of nature – the conservation laws – it is necessary to idealize, and refer to systems of a kind which never actually occur.

What applies to the elementary and most fundamental processes of nature applies also to more complex natural processes. The natural laws governing these processes derive from their intrinsic or essential natures. That is, they are laws concerned with the dispositional properties and structures of the natural kinds involved and the natural processes that these properties and structures necessarily generate. Therefore, to state the laws concerning natural processes, it will generally be necessary to idealize in order to exclude extraneous forces and influences. Natural processes rarely, if ever, occur in the actual world without some overlay of such influences.

Idealization therefore has a fundamental role in physical theory. It is an absolutely necessary device for conceptually isolating the causal processes that are the main subject matter of our inquiries. Without it, none of the laws of nature concerning causal processes could be stated. For these laws concern the intrinsic natures of these processes, and to say what the intrinsic nature of a process is, we have to be able to say how it would occur in the absence of disturbing influences, i.e., ideally.

6.13 THE DEMARCATION PROBLEM

The essentialist theory of laws of nature also solves the demarcation problem mentioned in Section 6.2 – the problem of distinguishing between formal principles and laws of nature. The problem arises in this way: If formal principles, such as those involved in neoclassical equilibrium theory in economics, and genuine laws of nature, such as the laws of electromagnetism, are both necessary, in the sense of being true in all possible worlds, then how are they different? What justifies us in calling one a formal principle and the other a more imposing law of nature? Conventionalists can consistently maintain that they are not fundamentally different, since they are both true by definition of convention. But most philosophers would wish to distinguish between them. However important economic theory may be, hardly anyone who is not a conventionalist thinks that it is dealing with laws of nature. Humeans can explain the difference between neoclassical economic theory and the theory of electromagnetism by saying that the laws of electromagnetism are empirical generalizations that just happen to be true in our world, whereas the principles of neoclassical equilibrium theory are true only in theory. Rational economic people would have to behave in a free market economy with perfect competition in the way that they are postulated to behave because this is how the key terms in equilibrium theory are defined. But a Humean then has a serious problem in explaining the kind of necessity that attaches to laws of nature. Are they really just generalizations that happen (so far) to have been borne out by experience? Or are they too just true in theory?

Scientific essentialism explains the difference by pointing to the different grounds of necessity. The principles of neoclassical equilibrium theory are true by definition or convention, and are therefore true only in theory. In essentialist parlance, they are necessary *de dicto*. The laws of electromagnetism, on the other hand, must hold of electromagnetism in any world in which electromagnetic radiation may exist. The laws had to be discovered empirically, of course, so they are *a posteriori,* in the way that all empirical generalizations are. But what has been discovered is the essential nature of such radiation – that is, the properties and structure that any radiation must have if it is to be electromagnetic radiation. The laws of electromagnetism are thus necessary *de re.*

The essentialist theory also explains clearly why there are strictly speaking no laws of nature in any of the social sciences. It is not because these sciences are in their infancy, or yet to mature fully, as many have suggested. It is because these studies do not concern natural kinds of objects, properties, or processes. If one is not dealing with natural kinds of things, one cannot then expect to discover processes of the kinds described by laws of nature. The social sciences are not immature; they are just concerned with a different kind of subject matter.

1. Hall (1954, 172) says: "... the concept of natural law in the social and moral senses familiar to medieval jurists, and (its employment in the phrase 'laws of nature') signifies a notable departure from the Greek attitude to nature. The use of the word 'law' in such contexts would have been unintelligible in antiquity, whereas the Hebraic and Christian belief in a deity who was at once Creator and Law-giver rendered it valid."
2. See Axiom 11 in Section 2.10.
3. See Armstrong (1983, 138).
4. See Section 7.3 for a development of a concept of scope that is applicable to laws of nature.
5. I use the term "causal laws" broadly to refer to the laws concerning the interactions and spontaneous behavior (for example, radioactive decay) of specific kinds of substances. I am aware, however, that not all laws of this kind are causal in the narrower, and perhaps more common, sense of this word.
6. For instance, in van Fraassen (1989).

7. With apologies to Shakespeare. "...some men are born great, some achieve greatness, and some have greatness thrust upon them" (*Twelfth Night*, II, iv, 158).

8. Cartwright (1983) distinguishes between phenomenological and theoretical laws. The theoretical laws, she argues, are neither true nor false, but propositions of an abstract theoretical system that has to be applied to tell us anything factual about the world. Chalmers (1987) objects to Cartwright's account, arguing that theoretical laws are about the basic generative mechanisms, structures, and powers that exist in the world, and so are consistent with realism in physics. Suchting (1988) argues that it does little to argue that a philosophical position is consistent with realism in physics. What has to be shown, if one wishes to defend realism, is that realism is the preferable position (p. 78).

9. See, for example, Scriven (1961).

10. Whether this theory is really the one that was held by Hume is now in some doubt. See Costa (1989), who argues that "Hume does not reduce causation to regularity in any direct or simple way," and that there is a case for saying that Hume was a causal realist, despite his insistence that nothing more than regularity is truly conceivable by us. However, this is a debate that need not concern us. The "Humean" theory of laws is well known and widely accepted, even if Hume himself did not embrace it.

11. Against this, Fales (1990) argues, persuasively in my view, that forces of the sort required to explain causal connections are conceivable by us. Our concept of causation, he says, derives from our experience of bodily force. (pp. 11–25). Thus, Fales confronts, head on, Hume's "inconceivability" objection to the claim that there are causal powers in nature (pp. 25–39).

12. This is the problem Bhaskar (1978) focused on in his critique of the Humean theory of science.

13. See Ellis (1965) for an alternative to Newtonian dynamics based on a different principle of natural motion, and Winnie (1970) for an alternative to Special Relativity that rejects the "convention" that the speed of light is the same in all directions.

14. In Bigelow, Ellis, and Lierse (1992).

15. In the sense defined in Fox (1987).

16. We think that the global laws are also grounded in the properties of things. But for these laws the things are worlds and the properties are global dispositions. In Bigelow, Ellis, and Lierse (1992), we argue that the world itself is an instance of a natural kind – one that is distinguished essentially from other kinds of worlds by its global properties and structures. If this is so, then the laws that hold in our world must hold in all worlds of the same natural kind as ours.

17. In this respect, our theory is like Fales' (1990).

18. Huw Price (1999) has argued that the hypothesis of non-locality can be avoided if a theory that accepts the possibility of backward causation is accepted.

19. This is the most serious objection to the scientific essentialist account of natural necessity I am aware of. It is further discussed in Sections 7.8 and 7.11.

20. But see the discussion in Section 7.5.

21. It is of course contingent what individual thing happens to be in the position of being observed by us. Something else might be, or might have been, substituted for it. But that is another matter.

22. Fales (1986) claims that the essential properties of natural kinds are just their

monadic properties. However, I think we must recognize that there are compound natural kinds, and that even elementary natural kinds (such as electromagnetic fields) may have internal structures.

23. One of the best discussions of the relationship between natural kinds and the real essences of natural kinds is to be found in Brody (1967).

24. The qualification "in worlds like ours" derives from the hypothesis defended in Bigelow, Ellis, and Lierse (1992) that the universe itself is an instance of a natural kind, and that the laws of nature are of its essence. A similar thesis was defended in Harré and Madden (1975).

25. Van der Waals' gases, for example, are certainly more realistic than perfect gases, because the theory takes account of the volumes occupied by the gas molecules and of the fact that gas molecules do not move independently of each other. But everyone knew that the molecules of a gas were not just point masses, and that they interacted with each other. So Van der Waals' theory did not add much to our theoretical understanding of gases or of the thermodynamic processes involving them. What it did was show how to make the abstract dynamical model more realistic, and therefore more useful for practical purposes. Van der Waals' theory is therefore of considerable interest to engineers who have to deal with real gases.

7

Natural Necessity

7.1 THE PROBLEM OF NATURAL NECESSITY

The two most important problems associated with laws of nature are undoubtedly the Necessity and Ontological Problems. Most philosophers are likely to find the essentialist theories of idealization and demarcation outlined in the last chapter fairly satisfying, provided that a plausible essentialist theory of natural necessity can be developed and a good solution to the Ontological Problem can be presented. The focus of this chapter will be on the concept of natural necessity. The all-important Ontological Problem will be dealt with in the next chapter.

7.2. HUME'S CONCEPT OF NATURAL NECESSITY

Hume's treatment of the problem of natural necessity dealt mainly with the relationship between cause and effect. He argued that all reasoning concerning matters of fact is ultimately founded on this relationship. So, he thought, the more general problem of justifying all sound reasoning of this kind could be solved if the more specific one of justifying reasoning from cause to effect could be. What then, he asked, is there about this relationship to justify such an inference? Is there, perhaps, some kind of necessary connection between causes and effects? On this question we have conflicting intuitions. On the one hand, it seems obvious that the things we do and the events that occur in nature have effects, and that these effects are somehow produced, or brought about, by these actions and events. The effects would seem to be not just subsequent happenings but, in most cases, inevitable consequences of the actions or events that give rise to them. On the other hand, the effects that are produced do not seem to be necessitated by their causes in the strong logical sense in which, say, adding one to an even number produces an odd number. For the contrary of any cause and effect relation is easily imaginable, "as if ever so conformable to reality." Therefore, if causes necessitate their effects, then they must do so in some weaker, non-logical, sense.

But then, Hume asked, what is the foundation in experience of this idea of necessitation, or necessary connection, between events? What is its source? He examined a number of specific cases of causation in a mock attempt to find it, and concluded that there is nothing whatever in any of these cases, considered one at a time, to suggest the idea of necessary connection.[1] So, in the following very famous passage, he concluded:

Upon the whole, there appears not, throughout all nature, any one instance of connexion which is conceivable by us. All events seem entirely loose and separate. One event follows another; but we never can observe any tie between them. They seem *conjoined,* but never *connected.* And as we can have no idea of any thing which never appeared to our outward sense or inward sentiment, the necessary conclusion *seems* to be that we have no idea of connexion or power at all, and that these words are absolutely without any meaning, when employed either in philosophical reasonings or common life (Hume, 1777, 74. Hume's emphases).

He went on to say that there is a source of the idea of necessary connection – namely, in the effects on the mind of repetition. Repeated observations of instances of causal sequences produce associations of ideas which determine our expectations about any new cases we may come across. But in the world itself, the world we are observing, Hume argued, there are just regularities of various kinds.

Nearly everyone agrees that Hume's account of causal necessity will not do, and I do not propose to add further to the literature on this. But the problem remains: How can the conflict of our intuitions about the relation of cause to effect be resolved? Is Hume right in thinking that there are no necessary connections between causes and effects? If so, then how can the illusion that causes somehow necessitate their effects be accounted for? Is there some other, and better, way than Hume's of accounting for the appearance of causal necessity, while denying its reality? If, on the other hand, there really are necessary connections between causes and effects, then what is their nature? Are they real or metaphysical necessities, as I believe? Or is there perhaps a weaker kind of necessary connection holding between causes and effects, as many writers have supposed – one that is compatible with the contingency of causal laws?

Hume's argument to the conclusion that there are no necessary connections between events depends on his thesis that if a contrary of a given proposition is imaginable, and there is no sound argument to show that it is impossible, then it is indeed possible, and only experience can teach us that it is not so. This is the argument that Hume uses over and over again in both the Treatise and the Enquiries to establish his conclusion that there are no necessary connections in nature. Let us call it Hume's "imaginability test" for logical possibility.

7.3. IMAGINED POSSIBILITY

Using this test, Hume was able to argue persuasively that causal laws are not necessary since their contraries are always imaginable, and the arguments from experience that these contraries are physically impossible are either question-begging or depend on assumptions that are as much in need of justification as the conclusion to be justified. So successful was Hume's strategy of argument that his conclusion that causal laws are contingent and have the status of universal generalizations based on experience has been, until very recently, generally accepted by philosophers everywhere in the Anglo-American tradition. Nevertheless, I think that the argument is unsound, and in this chapter I shall take up the challenge it presents. I shall argue that what is imaginable is quite different from what is really possible. The imaginability test for possibility does not pick out any ontologically significant class of events.

While many philosophers clearly welcome Hume's conclusion that events are not necessarily connected, there are other implications of his imaginability test that may not be quite so welcome. For the same argument might be used to show that there are also no necessary connections between the spatial parts or temporal stages of things, even where they are physically inseparable. Consequently, we must conclude, if we are consistent Humeans, that everything, and every part and stage of everything, is unconnected with everything else. The conclusion of this line of argument would be as follows:

Upon the whole, there appears not, throughout all nature, any one instance of connexion which is conceivable by us. All of the parts and temporal stages of things seem entirely loose and separate. One part or stage of a thing abuts or immediately succeeds another; but I never can observe any tie between them. They seem *conjoined*, but never *connected*. (My parody of Hume.)

Probably, Hume himself would have had no problem with this conclusion because he would have rejected all cohesive forces just as firmly as he rejected the dynamical ones. But at least some of his followers would be embarrassed by it. For the parts and stages of objects do not always seem to be quite as loose and separate as distinct events might be supposed to be.

One can easily imagine a bronze statue just dissolving into a heap of dust. If this were really to happen, it would be astonishing, but probably most philosophers would agree that it is logically possible. One can also imagine the same statue growing wings like a butterfly and fluttering away, or singing Christmas carols, or getting up and walking through a brick wall. Are these things all logically possible too? Probably most philosophers would say that they are, since there are no evident contradictions involved in supposing that they happen. "Of course," they would say, "such things could not possibly happen in the actual world." But they are imaginable, and they would argue that this is enough to show that they are possible. Should I concede, then, that everything that is imaginable is possible? I think not.

Imaginability is a very bad test of possibility. True, there are no surface contradictions involved in the descriptions of these imagined happenings. But this is not reason enough to think that such happenings are possible? It would be, perhaps, if anything could behave in any way at all, as the holy fathers of the Council of Trent believed (see Section 7.7). But it is not enough if the identity of a thing depends on how it is constituted, and so ultimately on the causal powers, capacities, and propensities of its constituents.

The imaginability test of logical possibility derives from the assumption that what a thing can do or become depends only on its manifest image. For this is what the imagination has to work with. It starts with the manifest image and transforms it. But why should we suppose that all imaginable transformations are possible? In supposing this, the imaginability test operates in the wrong way and at the wrong level. It assumes bizarre phenomenological views of change, and of identity through change. For the test assumes that if the manifest image of something can be transformed by degrees into that of something else (as it nearly always can be), then this transformation is logically possible for the things themselves. However, what is really possible for a given thing to do or become does not depend only on the transformability of its manifest image. It depends also on what kind of thing it is, and how and of what it is constituted. A horse cannot, really cannot, be transformed into a cow, although an image of a horse can (easily with modern technology) be transformed into an image of a cow.

In considering questions of possibility, it is important to keep the distinction clearly in mind between what a thing is and what it looks like.

There might conceivably be a creature in some possible world that looks like a horse, which can indeed be transformed into something that looks like a cow. But it could not possibly be a horse since horses are incapable of any such transformations. Nor could the result of the transformation be a cow because cows cannot be produced this way. It is like that stuff XYZ on "twin earth." It may look like water, it may be called "water," and twin earth "people" may actually use it like water. But it cannot really be water if it is differently constituted, and so does not behave chemically as water does. What a thing can do or become depends on the kind of thing it is. It does not depend on what it looks like, or what kind of thing we think it is. What we think is a horse might conceivably not be a horse. Therefore it is epistemically possible that we are mistaken in thinking that something is a horse. Therefore it is epistemically possible that the creature before us, which we take to be a horse, is not a horse. Therefore it is epistemically possible that it could behave in ways that horses could not possibly behave. Therefore if we can also be mistaken about the kind of world we inhabit (which of course we can), we must admit that it is epistemically possible that the thing in front of us could be transformed into something that looks like a cow. But if the object in front of us is really a horse, then it cannot be so transformed, because it is not possible for a horse to become a cow, or even to come to look like a cow. A horse could no more become a cow than it could become a banana. Nor could a horse even come to look like a banana, or any other kind of thing it does not already closely resemble.

The imaginability test of possibility thus confuses what is really possible with what is only epistemically possible. It purports to be a test of what could, really could, occur in some given circumstances, when in fact it tells us only what we are able consistently to imagine happening to things that are superficially like those that exist in these (or in superficially similar) circumstances. The two concepts of possibility – epistemic and real – may indeed cut across one another. For not only might what is epistemically possible be really impossible, the converse might also be true. That is, what is epistemically impossible might really be possible. For what we are able to imagine is presumably conditioned, and hence limited, by our common experience of middle-sized particulars. Consequently, what we may be able to imagine is unlikely to tell us much about what can really exist or occur at the truly macroscopic or microscopic levels. If no process by which a certain quantum effect can be produced can be imagined, it does not follow that no such process is possible.

But what about logical possibility? It would be widely agreed that imaginability is not a sufficient condition for physical possibility, and so not a sufficient condition for what I am calling "real possibility." But surely it is sufficient for logical possibility. After all, it might be argued, I might be quite wrong about what the laws of nature are, and hence wrong about what is physically possible. But if we can imagine something occurring, then there cannot be anything logically impossible about it. For we cannot imagine contradictions. Is there, perhaps, some other account of natural necessity compatible with the supposed contingency of the laws of nature? Must I, in rejecting Hume's theory of natural necessity, also reject Hume's test of logical possibility?

7.4 REAL AND LOGICAL POSSIBILITY

Real possibilites are the correlatives of real necessities, and these in turn are a species of logical necessities. However, real necessities are different from *formal* logical necessities – which are true in virtue of their logical forms – and *analytic* necessities – which are true in virtue of the meanings of words. Real necessities are a species of logical necessities because, like all other necessary truths, they are true in all possible worlds. Correlatively, something is really possible if and only if it is true in some possible world. The main difference, it would seem, is just that real necessities and possibilities are *a posteriori*, whereas logical necessity and possibility claims are *a priori*.

Real necessities depend on subject-matter in a way in which formal logical necessities do not. Formal logical necessities are not only true in all possible worlds, they are true under all interpretations of their nonlogical terms. This is not the case for real or metaphysical necessities; they are not true under all interpretations of their non-logical terms. However, this is not an adequate characterization of the difference between these two modalities, because analytic statements are also distinct from formal logical necessities in this way. They are necessarily true, but they are not true under all interpretations of their non-logical terms. How then should the different kinds of necessity be distinguished, and what are the important differences between them?[2]

The differences have to do with the grounding of these modalities. According to the theory developed in *Rational Belief Systems* (Ellis, 1979), formal logical necessities and possibilities derive from the laws governing the structure of ideally rational belief systems, and are reflected formally in the languages we use to describe the world. For most philosophers, however, formal logic is not part of the theory of rationality; it belongs to the theory of truth preservation, and the laws of logic are the laws of truth.

For the purposes of this book, it does not matter much which of these two accounts of logic is accepted. Natural necessities and analytic propositions are clearly not formal logical truths. If they have anything in common with them, it is that they are necessary in the same generic sense – that is, they are true in all possible worlds. The differences between them lie in the reasons why this is so.

To discover what is formally logically possible, we never need to investigate the subject-matter of our discourse. For our aim is not to discover anything about this subject matter. From my point of view, it is simply to discover what kinds of propositions and inferences we can or cannot rationally deny, given only an understanding of the connectives and operators of the formal language we are studying. From the point of view of a Fregean theorist, the aim of formal logical inquiry is to discover what propositions or inferences must be true or formally valid, given only an understanding of the laws of truth or truth preservation for the connectives and operators of the language. Either way, these concepts of necessity and possibility depend on logical form alone.

The more interesting distinction is between what is analytic and what is metaphysically necessary. For these kinds of necessity are not purely formal; they depend either on the meanings of the non-logical terms of the language, or on the natures of the objects (properties, processes, and so on) to which reference is made. Roughly, the distinction is this: Analytic propositions are true in virtue of the meanings of words – that is, they depend for their truth on some conventionally established criterion for including something in some linguistically defined class. Metaphysically necessary propositions, on the other hand, are true in virtue of the essential natures of things – for example, they state correctly, or otherwise depend for their truth on, what makes something a thing of the natural kind it is. Analytic statements are thus true just because our linguistic conventions are as they are. Metaphysically necessary propositions of the kind in which we are interested are true because the world is divided into natural kinds (distinguished categorically from each other in the way that natural kinds are on the basis of their intrinsic properties and structures) in the way that it is.

To judge whether a necessary proposition about some class of objects is analytic or metaphysically necessary, one technique is to abstract from the descriptive language used to refer to this class, and replace the general name used with an ostensive "kind-referring" expression, such as "stuff of this kind" or "things of this kind." If the necessity survives this process, then we know that it cannot be grounded in the descriptive language we had been using. "Water is H_2O," for example, clearly survives this test, because "stuff of this kind is H_2O," said pointing to a glass of water, is no less necessary than "water is H_2O." If there is any doubt about it, then it can only be a doubt about what the intended object of reference is, or ignorance about what its nature is. Once it is known that (a) it is the stuff in the glass that is being referred to, not, for example, to the glass itself; (b) this stuff is H_2O; and (c) the physical and chemical properties of stuff like this all stem from the fact of its being H_2O, then it is clear that the proposition is necessary. The stuff is H_2O, and it could not be, have been, or become anything other than H_2O.

Contrast "Water is H_2O" with "a bachelor is an unmarried man." Replace "a bachelor" with "a person of this kind" said with reference to someone who happens to be a bachelor. "What kind?" one wants to ask. Does this individual represent a class of individuals that is naturally distinct from any other class of individuals, so that reference to any one member of the class is sufficient to establish reference to the whole class? I think not. Examine any given bachelor as thoroughly as you please; you will never discover the intended reference of the word "bachelor" as a result of such an examination. For bachelors are not bachelors in virtue of their intrinsic properties or constitutions – they are not intrinsically distinct from any other men. Hence the expression "a person of this kind," which is evidently intended to fix reference to a certain class, must fail to do so, unless some other means of identifying the class is provided.

There is an important distinction between what is possible for any given individual or kind and what is analytically not self-contradictory. Consider, for example, the proposition that Robin is an unmarried person who is not a bachelor. Probably most philosophers would say at once that this proposition describes a logically possible state of affairs, because Robin could be a woman. This is true: It does describe a logically possible state of affairs. But the reason why this is so is not obvious; and it is not any of the ones that would usually be given. Suppose that "Robin" actually names a man. Then to say that this proposition describes a possible state of affairs is to imply that Robin, who is actually a man, could have been a woman. But what does this mean? Of course, you could be mistaken in thinking that Robin is a man, when in fact she is a woman. But we are assuming that Robin is actually a man. So this possibility, which, in any case, is only an epistemic possibility, cannot be the possibility you have in mind when you say that Robin could have been a woman. You could, perhaps, be saying that "Robin" is as much a woman's name as a man's. Hence, "Robin" could refer to a woman. But we are supposing that in the context, "Robin" does not refer to a woman but actually refers to a man. How then could it have been the case that Robin is a woman?

Consider the following scenario. Robin had had his name down for a sex-change operation, but somehow he had missed his appointment with the surgeon. Since these operations are generally very successful, we can say quite confidently, that, but for the missed appointment, Robin would now be a woman. But then, Robin would not now be a bachelor. He would, however, have remained unmarried. Therefore, if it were not for the missed appointment, Robin would now be an unmarried person who is a not a bachelor. Therefore, even though Robin is in fact a man, the state of affairs described by the sentence "Robin is an unmarried person who is not a bachelor" clearly could have been realized. Therefore the judgment that this sentence describes a possible state of affairs is correct. But note: This judgment depends on the assumption that Robin's identity is capable of surviving a sex-change operation, and so does not depend only on the formal structure of the language, or on the intentions of the terms used to describe Robin's marital status.

The dependence of real necessities and possibilities on facts about the identities of the objects of reference is mainly what distinguishes these particular modalities from other full strength modalities. For these facts cannot be ascertained just by reflection on the language we use to describe the world. The judgments we make about what is really possible must also depend on what kinds of things we take the objects of reference to be and what we think their essential natures are. Hence our judgments concerning what is really necessary or possible must depend on our theories about the identities and intrinsic natures of the things and kinds of things we may refer to. And if the required theories about these things are *a posteriori*, then the judgments we make of what it is necessary or possible for them to be or become must also be *a posteriori*.

7.5 INDIVIDUAL ESSENCES AND KIND ESSENCES

Some individuals have intrinsic properties that can change over time. Metals can suffer fatigue, structures can become altered, things can become more elastic, people can acquire beliefs they did not have before, their skills can improve or deteriorate, their concerns for their fellows can diminish, and so on. So, not all individuals, qua individuals, hold all of their intrinsic dispositional or structural properties necessarily. Some intrinsic properties are held contingently and therefore accidentally. However, there are some things – those that belong to natural kinds – that hold some or all of their intrinsic properties necessarily in the sense that they could not lose any of these properties without ceasing to be things of the kinds they are, and nothing could acquire any set of kind-identifying properties without becoming a thing of this kind. These kind-identifying sets of intrinsic properties are the ones I call "the real essences of the natural kinds," although they might perhaps more correctly be designated the "kind-essences" of the individuals that possess them, since the bearers of these properties are undoubtedly the individuals.

Besides having a kind-identity, an individual thing may also have certain properties in virtue of which it is the individual it is. That is, it may have certain properties that it could not lose without ceasing to exist, and that nothing could have acquired, except in the process of becoming that very individual. Often, when people speak of the real essence of a thing, it is the individual essence they have in mind. Saul Kripke, for example, talks of individual essences rather than kind-essences (Kripke, 1972). Mostly it is assumed that the individual essence of a thing belonging to a natural kind includes its kind-essence. That is, the identity of an individual is supposed to depend on its being just the natural kind of thing it is. If this is right, then an individual of one kind could not possibly be transformed into something of another kind, although it might cease to exist and be replaced by something else. However, I am not at all sure that this is right.

Certainly, there are limits to the possibilities of inter-specific transformations. A horse could not become a cow, for example, because there is no natural kind of process, or complex of natural kinds of processes, by which such a transformation could occur. But when a change of nature occurs as a result of a natural kind of process, it is not at all clear that the individual undergoing such a transformation could not survive this process. For example, if an atom loses an electron by β-emission to become an atom one greater in atomic number, it is not obvious, nor is it even plausible, that the former atom has just ceased to exist or that a new atom has come into being at precisely the place where the first atom was. Most plausibly, the former atom has just lost a nuclear electron, and thereby changed its nature. For there is a powerful continuity argument to suggest that as an individual, the former atom still exists, but now as an atom of another kind.

For these reasons, I am reluctant to accept, as Bigelow (1999) urges me to, that the individual essence of a thing belonging to a natural kind includes its kind-essence. Nor do I think that I have to accept this thesis to provide a sound basis for scientific essentialism. For individual essences

238

would seem to have very little to do with kind-essences. The identity of something as an individual seems to depend primarily on its temporal and causal history, and therefore on its extrinsic, not its intrinsic, properties. Therefore a separate argument would be needed to show that an individual cannot change its kind essence. If, as a matter of fact, individual things cannot ever change from being essentially a thing of one kind to essentially something of another kind, or cannot do so except within very narrow limits, then this is presumably a fact about the kinds of things that exist in our world, rather than a necessary condition for individual identity.

Be that as it may, it is not necessary for me to suppose that the individual essence of a thing belonging to a natural kind necessarily includes its kind-essence. For the theory of natural necessity I propose does not depend on it. To deduce the causal laws we need to know only what the kind-essences of the various natural kinds of things are, and what the essential properties of the properties that constitute these kind-essences are. For remember that the properties that constitute the kind-essences are intrinsic properties of the things that have them. And these intrinsic properties must include all of those intrinsic dispositional properties of the things of these kinds possessed they have solely in virtue of being things of these kinds. Therefore the intrinsic dispositions of these things to interact with each other must be entailed by the fact of their being the kinds of things they are. Now, the causal laws concerning two natural kinds of things A and B are descriptions of ways in which things of the kinds A and B are intrinsically disposed to interact with each other in virtue of their being things of the kinds they are. Of course, as Bigelow points out, things may not interact as they are intrinsically disposed to interact. For other forces may come into play. But then the laws of nature that we call "causal laws" allow for this. The causal laws are not contingent universal generalizations about how things actually behave, but necessary truths about how they are intrinsically disposed to behave.

Bigelow's mistake is to think that if the laws of nature do not derive from the individual essences of things, then they cannot be necessary *de re*. At best, they can only be necessary *de dicto,* if indeed they are necessary at all. But this is not so. The laws of nature that I call causal laws are all necessary *de re*. The laws in question are straightforward descriptions of the essential properties of the intrinsic dispositional properties that fundamental things must have in virtue of being things of the kinds they are. Let K_1 and K_2 be natural kinds, "$\Rightarrow$" be the connective "if . . . then . . .", "C_i" the predicate "are in the specific circumstances C_i within a range of possible circumstances C," and "IE_i" the predicate "are intrinsically dis-

239

posed to interact in the manner E_i within the range of possibilities E."
Then the general form of a causal law involving two natural kinds of things
K_1 and K_2 is:

L1. For all x,y and i, necessarily $[x \in K_1 \ \& \ y \in K_2 \Rightarrow (C_i x,y \Rightarrow IE_i x,y)]$

For example, let K_1 be the kind that consists of samples of pure hydrogen,
K_2 the kind that consists of samples of pure oxygen, and i an index of circumstances in which such pure samples of hydrogen and oxygen are disposed to interact chemically. Then the causal law would be something like:

For all x and y, it is necessarily true that if x is oxygen and y is hydrogen, then if
x is ignited in y, then x and y are intrinsically disposed to combine chemically to
form water.

Note that the necessity operator that occurs in a law such as this, which
spells out the the intrinsic dispositional properties of the kind-essences of
things, occurs within the scopes of the universal quantifiers – that is, in a
de re position. If, as Bigelow insists, individual essences include kind-
essences, then the causal law is certainly necessary *de re*. On Bigelow's thesis what we have is:

L2. For all x,y and i, $[x \in K_1 \ \& \ y \in K_2 \Rightarrow$ necessarily $(C_i x,y \Rightarrow IE_i x,y)]$

Consequently, if we knew that a $\in K_1$ & b $\in K_2$, we could instantiate to a
and b and detach the consequent to obtain "necessarily $(C_i a,b \Rightarrow IE_i a,b)$."
However, such a strong essentialist claim entails that individuals that belong to natural kinds cannot in any circumstances change their natures.
This is the thesis which Bigelow urges me strongly to accept. However,
for the reasons given, I am not sure that I have to, or that I want to. I remain to be convinced that I must.

The weaker thesis, represented by **L1,** has one clear advantage over
L2. If **L1** is accepted, then it will be accepted on the basis that causal laws
derive not from the essential properties of individuals, but from the natures of the intrinsic properties that individuals belonging to natural kinds
must have in virtue of being members of these kinds. The beauty of this
is that it allows us to generalize the account of natural necessity and to explain the necessity of those laws of nature that can be stated without reference to natural kinds of objects.

In Chapter 2, we described an ontology that includes natural kinds of properties as well as natural kinds of objects. Gravitational mass, for example, is a generic kind of intrinsic causal power. The infimic species of this generic kind are the specific gravitational masses that individuals may possess. Let M_1 and M_2 be the kinds consisting individual objects of masses m_1 and m_2 respectively, and $r(x,y)$ be the proposition that x and y are separated by a distance r. Then, analogously to **L1**, we have:

M1. For all x,y and r, necessarily $[x \in M_1 \ \& \ y \in M_2 \Rightarrow (r(x,y) \Rightarrow IE_r(x,y))]$

where $IE_r(x,y)$ is the statement that x and y are intrinsically disposed to accelerate toward each other with accelerations proportional to m_2 and m_1 respectively, and inversely proportional to r^2. However, the law analogous to **L2** – namely,

M2. For all x,y and r, $[x \in M_1 \ \& \ y \in M_2 \Rightarrow$ necessarily $(r(x,y) \Rightarrow IE_r(x,y))]$

is fairly evidently untenable, unless x and y happen to belong to natural kinds. For **M2** requires that the members of M_1 and M_2 have their specific masses essentially. They may do so; but fairly clearly they need not do so.

7.6 "POSSIBLE WORLDS" THEORIES OF NATURAL NECESSITY

Scientific essentialists hold that one of the primary aims of science is to define the limits of the possible. That is, it contends that scientists seek, wherever possible, to discover what can or cannot happen, depending on the circumstances, and where something is found to be possible, to determine the probability of its happening. The case for this was argued in Chapter 4. It explains clearly the concern of physical natural scientists to discover the essential natures of things and the laws of action of their dispositional properties. It explains also science's concern with the construction of essentialist explanations.

The involvement of empirical theories in a great many of our judgments of necessity and possibility is obscured in most "possible worlds" theories of these modalities. For these theories have been developed with an eye to providing adequate semantics for formal languages with modals and conditionals, not for exploring the realms of the possible. Consequently, "possible worlds" theories focus on specifying truth conditions for propositions of these kinds. Strictly speaking, there is nothing wrong with this, and everything I want to say about *a posteriori* modalities of the kind we have been discussing can be said using the language of "possible worlds" semantics. But because our focus is not on language but on reality and real possibilities, it may be better to develop the theory of what is possible with reference to the possibilities of existence rather than truth.

Standardly, it is said that "$\Diamond p$" is true at a world W iff "p" is true at some possible world accessible from W. However, if we follow Lewis, then what we imagine to be true at another world accessible from W will be true of different things from those that exist in W. It will be true of what Lewis calls the "counterparts" of the things in W. But if we are interested in what might happen to, or have been true of, the things in the real world, the counterparts and the possible worlds in which they exist must be very carefully constructed. First, the accessible possible worlds must be such that the very same entities as those that we are concerned with might have existed in them. Therefore the possible worlds accessible from this world must be worlds of the same natural kind as ours (see Section 7.9), and hence the laws of nature and the basic ontologies must be the same in all of these worlds. Second, the counterparts of the things whose real possibilities we are considering must be essentially the same as those in the actual world. A counterpart that just looks like the actual thing but is differently constituted or has a different intrinsic nature does not tell us what it is possible for the actual thing to be or become.

The logic of real possibilities and necessities is just S5, the same as for the logical modalities, although the second and higher order modalities in expressions involving iterated modalities are likely to be straightforwardly logical, rather than metaphysical. The basic logic must be S5 because the accessibility relation for real possibilities links possible worlds of the same natural kind, which is an equivalence class. It must therefore be a relation that is symmetrical, transitive, and reflexive. It is not universally agreed, however, that this is the logic of such modalities. In their book *Science and Necessity,* John Bigelow and Robert Pargetter (1990) propose a theory of natural necessity that argues for a different logic. The theory they propose embraces a kind of possible worlds realism, and depends on the strong form of scientific realism outlined in Section 4.2. It is also a thoroughly Humean theory, which is worth examining, if only because it commits many of the errors I am seeking to highlight.

Bigelow and Pargetter do not believe in dispositional properties, except as supervenient on categorical properties and laws of nature. They embrace what Lewis calls "the Humean supervenience thesis" – "we could not have any difference in modal [including dispositional] properties, *unless* there were also differences in non-modal properties" (p. 175). But in

242

order to embrace this Humean thesis, they have to insist that very few of the most basic properties in nature are genuinely occurrent.

From the Humean supervenience thesis, it follows that there are no causal powers, capacities, or propensities existing as properties in their own right. For all of these properties are modal properties. According to Bigelow and Pargetter's theory, any differences in the causal powers, capacities, or propensities of things would have to be grounded in some underlying structural or other non-modal differences. However, most of the most fundamental properties in nature would appear to be dispositional. The gravitational mass of a body is its intrinsic power to generate gravitational forces. The inertial mass of a body is its intrinsic capacity to resist acceleration by forces. The charge of a body is its intrinsic power to generate electromagnetic forces. The half-life of a fundamental particle is a measure of its intrinsic instability, which depends on its various intrinsic decay propensities.

In Bigelow and Pargetter's ontology, the world is a structure whose only first-order properties are non-modal. These first-order properties are universals, which are the same in all possible worlds. If anything has a dispositional property, then this is a world-bound property, and so not a universal. It is a property it has in virtue of its categorical properties and the laws of nature, and the latter might well be different in other possible worlds. Since dispositional properties are not universals, they cannot be admitted to primary ontological status. The fundamental particles must therefore be distinguished ontologically from each other, not by their causal powers, capacities, and propensities, as they currently are in physics, but ultimately by their primary, presumably Lockean, qualities of shape, size, structure, and so on. The various dispositional properties of the fundamental particles, must, according to Bigelow and Pargetter's theory, depend ultimately on the intrinsic, but at present unknown, constitutions of these particles.

We draw attention to this point because their commitment to a Humean ontology of actual existence occurs at an important stage in their argument, and without it their theory of laws would be untenable. They admit (p. 174) that the correspondence theory of truth is compatible with the existence of modal primitives. They reject this possibility, however, because they think that "modal properties must be supervenient on non-modal properties" (p. 174). Consequently, they restrict their attention in what follows in their book to "theories which aim to characterize possible worlds without appeal to modal primitives" (p. 175).

This step is crucial because Bigelow and Pargetter require that there be a Hume world corresponding to any given world. Their theory, and their logic, of natural necessity both depend on this assumption. The Hume world vis à vis any given world is a world exactly like the given world, containing the same things with the same first-order properties, standing in the same first-order relations to each other, but without the causal connections, or other relations of natural necessitation that distinguish laws from accidental generalizations (p. 280). The Hume world corresponding to any given world is a lawless world. Every first-order thing that exists in our world is supposed to exist in the corresponding Hume world, and everything that happens in our world is supposed also to happen also in this other world, but not by necessity. What happens in the Hume world happens purely by chance. However, if the basic dispositional properties of the fundamental particles and fields are their essential properties, as I and others think,[3] then it is metaphysically impossible for such particles and fields to exist in a Hume world. But if there were no particles or fields like ours, then there could be no atomic or molecular structures like ours, and hence no DNA or other organic materials from which human or other living things could be constructed. Consequently, a Hume world would have to be a totally different kind of world. For it could not contain any of the first-order kinds of things that exist in our world. If a Hume world that would look superficially like ours to any transworld being (if there could be such a thing) could exist, then it would be intrinsically very different from the actual world. It would differ from the actual world all the way down to the most fundamental kinds of things that exist in these two worlds. This Hume world would be a globally counterfeit world containing none of the first-order kinds of things that go to make up this world.

One can take away the laws of nature, and leave every first-order thing in the actual world the same (except that everything that happens now happens only by chance) only if the laws of nature are not grounded in these things but imposed on them. If they are grounded in the causal powers, capacities, and propensities of the fundamental physical kinds that exist in the actual world, then one cannot take away the laws of nature without also taking away the kinds of things they are concerned with.[4] Belief in the existence of a Hume world therefore depends on the rejection of scientific essentialism. If an essentialist theory of causal and statistical laws is correct, and the dispositional properties of the fundamental physical kinds are the truth-makers for such laws, then it is logically impossible for there to be a Hume world containing things of the same kinds as those that exist in our world.

The failure of Bigelow and Pargetter to take essentialist theories of natural necessity seriously is their biggest mistake. Yet, strangely, their strong argument for realism about theoretical entities should have led them to do so. They are anti-reductionist about most theoretical entities, and happily embrace such things as possible worlds, sets, and propositions. Yet they baulk at dispositional and other modal properties, and so are committed to reducing the intrinsic dispositional properties of the fundamental natural kinds to categorical properties and laws of nature. Why? Why not abandon the Humean supervenience thesis and build natural necessities into the basic ontology of the world?

Bigelow and Pargetter offer very little argument in favor of the Humean supervenience thesis on which they rely, except that "it would guarantee the definability of modal properties in non-modal terms" (p. 175). It would do so, however, only if one were prepared to buy possible worlds realism. But no one would be willing to buy realism about possible worlds if the theory did not also guarantee the definability of modal properties in non-modal terms. This is the only thing in favor of possible worlds. So it's either/or – either one accepts both Humean supervenience and possible worlds realism, as Bigelow and Pargetter do, or one rejects them both, as I do, and seeks to ground causal modalities and nomological connections in basic dispositional properties.[5]

7.7 HUMEAN AND ESSENTIALIST PERSPECTIVES ON REALITY

The Humean metaphysic on which Bigelow and Pargetter rely depends ultimately on a medieval distinction between a substance and its species attributes. First, it implies that the causal powers and capacities of things are properties that depend on what the laws of nature happen to be. If the laws were different, they would have to say, then so must their causal powers be different. Second, since they suppose the laws of nature to be contingent, it must be contingent no matter how things are disposed to act or interact, and consequently impinge on our senses, or indeed on any instruments we might use to observe them. Consequently, in other possible worlds, things might seem to be very different from how they appear to us in this world. Conversely, things with very different structures, and made of different substances, might be disposed to behave in exactly the same way, if the laws of nature were sufficiently different.

The distinction, embraced by Humeans, between a substance and its species attributes was confirmed at the Council of Trent in 1551 in the doctrine of transubstantiation:

245

If anyone shall say that, in the most holy Eucharist, there remains the substance of bread and wine together with the body and blood of our Lord Jesus Christ; and shall deny that wonderful and singular conversion of the whole substance of the bread into the body, and of the whole substance of the wine into the blood, the species of bread and wine alone remaining, which conversion the Catholic Church most fittingly calls Transubstantiation, let him be anathema (Session 13, Canon 2. Quoted under "Transubstantiation" in *Microsoft Encarta,* 1994).

What is supposed to occur in the Eucharist is that the wine and bread literally become the blood and body of Jesus, even though all their species attributes (including all of their physical and chemical properties) remain just those of wine and bread. Same observable properties, different substances. Fine, if you think that what a thing is is logically independent of what it does. But madness if you think otherwise.

For scientific essentialists, this doctrine is bizarre. Of course the substances are still bread and wine! Nevertheless, many modern philosophers would argue that it is not logically impossible for transubstantiation to occur[6] since there is no formal contradiction involved in the supposition that it does. Therefore, they would say, it is at least logically possible that the bread and the wine should really become flesh and blood, despite all appearances to the contrary. However, it does not follow from the absence of any formal contradiction that there really could be a world in which transubstantiation occurs. For whether it is possible in this broader sense hinges on what the identity of a substance depends on. If it depends on its causal powers, capacities, and propensities, as scientific essentialists would say, then there is no possible world in which flesh and blood could behave in the bizarre way required by the doctrine of transubstantiation. It is only if the identity of a substance is supposed to be independent of anything that might be discovered about it by observation or experiment that the doctrine of transubstantiation makes any sense. On any other assumption, flesh and blood could not manifest itself as bread and wine, in this or in any other really possible world. They would have to cease to be flesh and blood in order to become bread and wine.

Although many, perhaps most, Humeans would firmly reject the doctrine of transubstantiation, they nevertheless continue, embarrassingly, to embrace its metaphysics. For every Humean believes that things exactly like those that exist in this world could exist in worlds in which the laws of nature were different, and so, presumably, even in worlds in which the laws of nature were very different. But how these things would behave, and therefore what effects they would have on us, and on our measuring instruments, would depend on what the laws of nature were in these other worlds. So Humeans cannot declare the doctrine of transubstantiation to be incoherent.

The alternative, and essentialist, view of the matter is that causal powers, capacities, and propensities exist in nature, and the identities of the various kinds of things that exist are dependent on them. What makes things flesh and blood is their physical and chemical properties. Flesh and blood cannot lack these properties, and have those of bread and wine instead. Conversely, if anything has the physical and chemical properties of bread or wine, then it is bread or wine, as the case may be. It is certainly not flesh or blood. Many of the physical properties of bread and wine are dispositional. For they depend on how the samples would respond to various tests. But, plausibly, the chemical constitutions of these substances are categorical properties. For although certain tests must be carried out in order to discover what the chemical constitutions of bread and wine are, the claim that they have these constitutions is not simply a summary way of stating the results of these tests. However, the claim that flesh and blood might have exactly the same physical and chemical properties as bread and wine, if the laws of nature were sufficiently different, presupposes that the constituents of flesh and blood might behave in exactly the way that the constituents of bread and wine actually behave. It presupposes, therefore, that the identities of the constituents of flesh and blood do not depend on their dispositions to behave physically or chemically as flesh and blood do.

On what, then, do the identities of the constituents of substances such as flesh and blood depend? Ultimately, they must depend on the identities of their constituents, whose identities must depend on those of their constituents, and so on. And if this line of dependence is pursued, we must either come to some basic constituents whose identities do not depend on anything else, or we must go on pursuing identity down through ever deeper levels. On present evidence, the categorical properties drop out before we get to the deepest levels – that is, the identities of the most basic kinds of things depend *only* on their dispositional properties. That being the case, it is metaphysically impossible for flesh and blood, constituted as they are, to behave as the doctrine of transubstantiation requires.

On the view taken here, the essential properties of things are not occult, but discoverable and measurable by the ordinary methods of science. There is nothing especially obscure or hidden about them. They can be found just by observing how the instances of the various natural kinds act on, and interact with, each other. Of course, to discover the effects of any particular property, it may be necessary to observe what happens in circumstances in which this property may reasonably be assumed to be the only relevant cause, or to observe it in circumstances from which we may abstract from the known effects of other causes. But this is not abnormal practice in science. On the contrary, it is standard experimental method to attempt to isolate what we wish to examine from causes whose effects are not the objects of investigation, or, where we cannot do this, to discount for the known effects of the causes we are unable to eliminate.

7.8 GROUNDING NATURAL NECESSITIES IN THE WORLD

It was argued in Section 4.2 that the Humean package – which consists of a Humean ontology, the Humean supervenience thesis, and "possible worlds" semantics – is unacceptable as a basis for a theory of natural necessities. If natural necessities are not grounded in the actual world, but solely in relations between causally distinct worlds, none of which contains any causal powers, then natural necessities cannot necessitate. What does it matter how the actual world is related to other possible worlds if all the possible worlds are in themselves non-modal? What then are the alternatives?

To ground natural necessities in the world, it is necessary to develop an ontology of things capable of sustaining causal and other modal relationships. To do this, it is necessary to ground the laws of nature somehow in the world. To deal with the bulk of the laws of nature – the causal and statistical laws that describe the powers and structures of things belonging natural kinds – it is sufficient to recognize that these things have their kind essences essentially. We have already seen how this works in particular cases. However, there are some natural necessities that do not seem to be grounded in the kind essences of specific kinds of things – namely, the most general laws of nature. Such laws apply to all things, or to all events and processes, and are in this sense "global."

To deal with this problem, it is reasonable to suppose that the ontological categories of objects, structures, and processes are themselves very general natural kinds, characterized by the basic ontologies of these categories, and some common dispositional and structural properties of the things in these categories. The global laws, such as the conservation laws and the global structural laws, might thus be grounded in these global natural kinds. Let O be the natural kind that includes all the objects or substances that do or can exist in the world, E the kind that includes all events and processes that do or can occur in the world, and S the kind that includes all properties or structures that do or can exist here. Then, given

248

these generic natural kinds, we might argue that the global laws are just descriptions of their essential natures. Thus we might say that every event of the kind E is intrinsically conservative of energy and of all the other conserved quantities. Or else we might say that all spatio-temporal structures that can exist are necessarily species of S and hence restricted by the global restrictions on S imposed by General Relativity. There is, however, another, and probably equivalent, hypothesis that I prefer because it is more unifying and easier to argue for – that the world itself is an instance of a natural kind. It is a world, I wish to argue, that is distinguished from worlds of all other kinds by its global structural, dispositional properties, and its basic ontology.

7.9 THE WORLD AS ONE OF A KIND

It is possible to say a great deal about the world as a whole. We can point to global structuring principles, universal processes of world evolution, general symmetries, a common ontological basis of reality, a single origin of the universe, and various universally conserved quantities. Given a knowledge of these, we can say a lot about what kinds of things can exist or occur in the world, or in any world with the same structure and global properties as ours. The world we live in is not an amorphous or disconnected world, it seems, but a highly integrated and coherent structure.

We can say fairly confidently, first, that ours is an expanding four-dimensional space-time world that is structured according to the principles of General Relativity. It is also evident that there is a global causal structure. For all of the events and processes occurring in the world (including the Big Bang and the process of inflation) consist ultimately of energy (and other conserved quantity) transmission processes and the instantaneous changes of state by which such processes are initiated or terminated. The discontinuous changes of state (the elementary events) include causal interactions between particles and events of decay from higher to lower energy states. The energy transmission processes (elementary processes) that provide the causal links between elementary events all have at least these features:

(a) They are initiated by elementary events, and are terminated only in such events.
(b) They are inertial in the sense that they do not require any external forces or other external causal mechanisms to sustain them.
(c) They are conservative of mass-energy, charge, spin, momentum, and all other universally conserved quantities.

249

(d) If they transmit information concerning such quantities, then they do so at speeds not greater than that of light.

(e) They are quantum-mechanically indeterminate – that is, they have no stages that are localized in space and time as events typically are.

These are, I would suppose, among the essential properties of all elementary processes.

If this basic ontology of events and processes is plausible, then the world's process must be just a complex of elementary causal processes. Now these processes are all stochastic in nature. That is, the times, places, and often the properties of the events that terminate them are not nomically determined by the times, places, and properties of the events that initiate them. This being the case, the steps between elementary events are like the steps in a random walk. There are definite probabilities of various outcomes, but the precise outcomes are never wholly predictable. It is reasonable therefore to think of the world as staggering, in the manner of a random walk, from elementary event to elementary event, and thus being systematically constructed in the sort of way in which a random walk is constructed. It is this process that constitutes the temporal evolution of the universe, and presumably of any world of the same natural kind as ours, which for our world constitutes time.

Second, the world is apparently one that displays certain global symmetries that are important for our understanding of what kinds of things and processes can occur. There is reason to believe that some of these symmetries may be broken from time to time. But even if the symmetry principles are not strictly universal, they are expressible as conditional probabilities in which the reference classes are universal – that is, as principles attributing genuine dispositional properties to all events and processes or to all physical systems. That is, they are global in scope.[7]

Third, the world is evidently a physical world – one that consists entirely of things that have energy and that interact with each other physically – that is, energetically. A physical world is not the only conceivable kind of world. Indeed, if dualist interactionism were true, then the world would not be a purely physical world.

Fourth, the world appears to be universally conservative with respect to a number of quantities, including energy, momentum, angular momentum, charge, and several others. Consequently, the world would seem to be one in which only certain kinds of changes are possible – those that are not forbidden by the conservation laws. Again, a caveat may be necessary. If some symmetries can sometimes be broken, then perhaps some of these quantities are not always strictly conserved. However, even if this is the case, the world's high degree of conservativeness with respect to these quantities is still a global fact about reality, and so, plausibly, characteristic of the kind of world in which we live.

Finally, the world is evidently one that is made up of a relatively small number of interrelated kinds of fundamental particles and fields, and it is plausible to suppose that when we know more about the basic structure of the world, we will be able to explain why these, and perhaps only these, particular kinds of particles and fields can exist. Certainly, if John Barrow, Frank Tipler, John Leslie, and other writers on cosmology are right, then the global properties of worlds of the kind in which we live are at least highly restrictive of the kinds of things that can exist in worlds like ours. It might even be the case that no things, other than things of the kinds that do exist in our world, could exist in any world with the same global properties.

Currently accepted theory may not, of course, be right in detail. But no one doubts that there are global properties and structure. And the fact that these global properties and this structure exist at all implies that the world is a unified whole. If the world consisted of unrelated kinds of things that were just thrown together somehow, these properties and this structure would be inexplicable. On the other hand, the existence of these properties and structure is sufficient to explain a great deal of what happens in the world. Certainly, they greatly restrict the range of possible kinds of things and occurrences. Therefore, we seem to be able to characterize the world, and explain why it behaves as it does, in much the same kind of way as we might characterize an electromagnetic field or a water molecule, and explain why it does the things that it does. Electromagnetic fields and water molecules are clearly instances of natural kinds. It is plausible, therefore, to suppose that the world itself is an instance of a natural kind. If one does not accept a theological theory of laws, it is hard to see how else, or how better, to explain the existence of all this global structure.

While not directly supporting the claim that the world is a member of a natural kind, Evan Fales (1990) supports a related thesis of the strong interconnectedness of universals, which he says implies that a world with different laws would be essentially different and would contain none of the same properties or kinds.[8] Given his interconnectedness thesis, Fales argues that:

We can imagine ... that some or all of the physical laws of our world did not obtain. We can even imagine counterfactually, there being sentient creatures like ourselves having under these circumstances sensations of kinds qualitatively indistinguishable from ours. But such a world would not contain the physical properties which our world contains. Because of the systematic ramification of causal connections, it would contain no identical physical laws. Transworld identity of properties cannot be separated from transworld identity of laws (p. 219).

Fales's strong interconnectedness thesis and our global natural kind hypothesis are not so very different. If all universals are strongly interconnected in the sort of way that he imagines, and laws are naturally necessary relations between universals, as he supposes, then the properties that exist in any world and the laws that govern their instances are very closely related: They come as a package. We think much the same. The natural kinds of objects that exist depend ontologically on their essential properties and structures. That is, these objects could not exist if these properties or structures did not exist, or were in any way different. The natural kinds of processes, including all primitive causal process, depend ontologically on what natural kinds of objects, properties, and structures they involve essentially. That is, these processes could not exist if these kinds of objects, properties, and structures did not exist. All objects and processes that do not belong to natural kinds depend ontologically on objects and processes that do, since those very same objects and processes could not exist, or occur, in any world in which any of the natural kinds of things of which it is constituted did not exist. Therefore the kinds of objects and processes that actually exist or occur could not exist or occur in any possible world except one with the same fundamental property universals and the same spatio-temporal-energy structural possibilities as ours. Therefore, if the world in which we live is a member of a natural kind, as we suppose, then its essential properties and structures will be those that any member of this kind must have by virtue of its being a member of this kind and that no member of the kind could lack. No member of the specific kind to which our world belongs could possibly lack any of the fundamental properties that exist in our world. Nor could it fail to have the spatio-temporal-structural possibilities that must exist if the structures that exist in fundamental objects or processes are to be possible. For the possibility of their existence is a necessary condition for the possibility of existence of the things that constitute our world. Therefore, any world of the same natural kind as ours must contain the same fundamental property universals and have the same spatio-temporal-structural possibilities as ours.

The hypothesis that the world is a member of a natural kind thus leads to the conclusion that the actual world is essentially one in which things have the same fundamental properties as those that exist in our world, and in which structures of elementary things having these properties are restricted or circumscribed by the global laws which hold in our world. The hypothesis that the world is a member of a natural kind enables us to explain what natural necessity is for our world, and, incidentally, for any other worlds of the same kind as ours. What is naturally *necessary* in our world is what must be true in any world of the same natural kind as ours. What is naturally *possible* is what might be true in a world of the same natural kind as ours. What is naturally *contingent* is what might or might not be true in a world that is essentially the same as ours.

7.10 ALIEN WORLDS AND ALIEN KINDS OF THINGS

The global kind hypothesis that has been put forward does not provide an adequate foundation for a general theory of natural necessity. For such a theory, we should need a general global kind hypothesis, according to which every real world is a member of a natural kind. With this general hypothesis, perhaps, natural necessity could be defined with corresponding generality. We could say, for example, that what is naturally necessary in a given world W is what must be true in any world of the same natural kind as W, that what is naturally possible in W is what might be true in a world that is essentially the same as W, and that what is naturally contingent in W is what might or might not be true in a world of the same natural kind as W. However, the general global kind hypothesis is more than is strictly needed to understand what natural necessity and possibility are for this world. It would only be required if it were thought that different kinds of things might be naturally necessary in other kinds of worlds, and the aim was to develop a general theory of natural necessity that would be applicable to these other kinds of worlds, as well as to our own world. However, I do not know how to argue convincingly for a general global kind hypothesis, except to say that it would be surprising if our kind of world were the only kind there is. But the logical possibility of there being worlds that have no global structure or fundamental properties, and hence no essential nature, cannot be ruled out.

It is plausible to suppose that some of the natural kinds existing in our world might not exist in other worlds that are essentially the same as ours. For example, there might be a world that is just like ours, but in which human beings do not exist. It is also plausible that the circumstances in which a given natural kind exists in our world might not be matched in other worlds of the same kind. In all such cases, it would be contingent

253

that these specific natural kinds exist in the world, or that they exist in the kinds of circumstances in which they do. Nevertheless, in any world in which these kinds of things do exist, they must have the intrinsic properties and structures that are essential to them. And if they exist in the circumstances envisaged, then they must behave as things of this kind have to behave in such circumstances. However, if it is contingent what specific natural kinds there are, or what kinds of circumstances exist, then it is also contingent what natural necessities will be displayed. If these specific kinds of things do not exist in a given world, then all we can say is that if they were to exist, then they would have to have these properties. If the envisaged circumstances did not occur, then we could only say that if they were to occur, then things of these kinds would have to behave in the manner dictated by their essential natures. The essentialist theory of natural necessity being proposed thus promises to yield an essentialist theory of natural necessitation as well. This issue will be discussed in the next chapter.

In considering the possibilities of other kinds of worlds, the questions naturally arise as to whether the kinds of things and processes that can occur in our kind of world could also occur in any other kind of world, and, conversely, whether things of kinds that could occur in worlds of other kinds could also occur in our world. If a kind A is a species of a kind B, then the essence of B must be included in the essence of A. Therefore, if the kind of world W_0 to which our world belongs is a species of a natural kind W, then the essence of W must be included in the essence of W_0. W must therefore be a general kind of world that may include many kinds of worlds besides those of our own kind. Our close cousins, if there are any, will presumably be similar in having at least some of the same global properties and structures as our world, thus permitting or restraining the possibilities of existence in these worlds in at least some of the same ways. Those worlds that are our siblings – other worlds in W_0 – must be supposed to have all of the same global properties and structures as our world and therefore precisely the same global laws and the same constraints on, and possibilities of, existence.

What can we say, then, about alien worlds? Not much. Presumably, the possibilities of existence will be enhanced or constrained by global laws in ways that are alien to us, and restricted or unconstrained by global laws that operate here. So alien worlds will generally contain alien kinds of things – things that could not exist in any of the worlds of W_0. And those things that exist in our own world will often be alien to the alien worlds, and so could not exist in them. Overlapping ontologies are possible, provided that our global laws are not so tightly interconnected that a change in any one of them would necessitate a change in all of the others.

If these speculations are correct, then they make for a nice theory of natural necessity. For, given these speculations, it is not metaphysically contingent what kinds of things, properties, or processes can occur in worlds like ours. Consequently, it is not metaphysically contingent what the laws of nature are in any such world. It might be epistemically contingent what kind of world we live in, because we might lack this knowledge, but alien natural kinds of things could not really exist in any world of the same natural kind as ours, and the same fundamental existents could not fail to exist in any such world. Therefore, although we may not know precisely what kind of world we live in, it is not metaphysically contingent what the laws of nature are for our world or for any world of the same natural kind as ours.

7.11 IS METAPHYSICAL NECESSITY TOO STRONG?

While many philosophers may be sympathetic to the idea that the laws of nature are in some sense necessary and immanent in the world, they are likely to baulk at the claim that they are metaphysically necessary. This seems far too strong. For it implies that the laws are true in all possible worlds, just as formal logical and analytic propositions are. But how can I possibly be in a position to assert such a thing? I claim, for example, that it is metaphysically necessary that every causal process be intrinsically conservative of energy and other universally conserved quantities. Why? Because, I say, it would not be a causal process of the global kind to which all events and processes occurring in this world belong if it were not conservative of these quantities. If a process were not conservative of these quantities, I say, then it would not be a process of a kind that could occur in our world, or in any other world of the same kind as ours. Bold claims indeed! But what evidence do I have for them? How do I know what kind of world we live in, or what kinds of objects, events, or structures are possible in our kind of world?

The fact is that I don't know and don't pretend to know. However, the conservation laws are well established, and what I think I can reasonably say is that if they are true, then they are necessarily true. They are necessarily true, if they are true at all, because what is true of the most general kinds of objects and events in this world must be true of the most general kinds of objects and events in all worlds of the same natural kind as ours. Such is the nature of natural kind reasoning.

255

Of course, the world might not be the kind of world we think it is. But this speculation trades on uncertainty, and is therefore of an epistemic nature. Of course, it is epistemically possible that the world is a very different kind of world from the kind we think it is. There might well be, for all we know, all sorts of monsters lurking in the dark, or in the future yet to be discovered, that violate these laws, in which case we may have to conclude that the world is truly, and perhaps necessarily, very different from the way we think it is. But epistemic possibility is not real possibility, and ignorance is not a source of knowledge. If you want to know what kind of world we actually live in, and therefore what is true of all worlds of the same natural kind as ours, you have to rely on the best theories available to you, and according to these theories, all events and processes are intrinsically conservative of a number of well-known quantities. I see no serious objection, therefore, to accepting the strong view that the laws of nature are all metaphysically necessary. The more specific laws of nature – the causal and structural laws – which depend on the causal powers and structures of the more specific kinds of things existing in the world, are undoubtedly metaphysically necessary. The most general laws of nature are laws of the same kind, I contend, but because we ourselves are world-bound, we cannot stand outside of the worlds to designate the natural kind of world to which our own world belongs. We can only designate it internally, postulate that it is one of a kind (for the sorts of reasons that have been elaborated), point to the kinds of objects, events, and structures that actually exist in our world, and try to say what their essential natures are. If we are right about all of these things, then what we assert to be true about the most general kinds of things existing in this world will be necessarily true.

7.12 FOUR PRINCIPLES OF NATURAL NECESSITY

1. Let w_0 be our world, and W the set of worlds similar to (of the same natural kind as) ours. Let $L_1, L_2, \ldots L_m$ be the set of global laws and principles that define the overall structure of worlds in W. Then, since these laws must hold in any world similar to ours, they are all naturally necessary in w_0. If w_i is any other world in W, then the laws $L_1, L_2, \ldots L_m$ are also necessary in w_i.
2. Let K_0 be the set of fundamental natural kinds of objects, properties, and processes in w_0. Then, by the assumptions I am making, these same natural kinds must all be instantiated in every member of W. That things of these kinds exist in w_0 is therefore necessary.
3. Let K_j be the set of ontologically fundamental natural kinds existing in any world other than w_0, but not in w_0. Then I make the supposition that no

member of K_j could be instantiated in any member of W. For if things of these kinds could exist in any member of W, it would have to be possible for them to exist in w_0. And since the members of K_j are supposed to be fundamental natural kinds, if they could exist, then they would have to do so. Therefore it is impossible for any such alien, but truly fundamental, kind of object, property, or process to occur in any member of W.

4. Let K_i be the set of natural kinds of objects, properties, and processes, other than the ontologically fundamental ones, that could occur in a world w_i that is a member of W. Then, necessarily, every member of K_i is an object, property, or process of a kind that could occur in w_0. Conversely, if K_0 is the set of natural kinds of objects, properties, and processes, other than the fundamental ones, that could occur in w_0, then, necessarily, these kinds of objects, properties, and processes are ones that could occur in any member of W.

It follows from 1 through 4 that whatever kind of object, property, or process can or must exist in any world of the same natural kind as ours can or must exist in our world, and conversely.

NOTES

1. Fales (1990, 11–39) argues that Hume is wrong here, even if we grant them his methodology. For the experience of bodily force is sufficient to ground the idea of there being necessary connections in nature.
2. In Kripke (1972, 304) the suggestion is made that "physical necessity *might* turn out to be necessity in the highest sense." Levin (1987) argues that this suggestion is untenable. He says: "If 'gravity' is rigid, the seemingly important consequence follows that gravity could not have been other than it is. Objects might have tended to accelerate toward each other as the inverse of the cube of their distance, and scientists in Newton's epistmological position could have discovered this, but they would not have inhabited a world in which *gravity* followed an inverse-cube law. Theirs would have been a world without gravity" (p. 290). But how does Levin know that objects "might" have tended to accelerate toward each other as the inverse of the cube of the distance? Is it not precisely because he thinks that the laws describing the dispositions of objects to accelerate towards each other, which the law of gravity was postulated to explain, are contingent? If so, then he begs the question against any essentialist who would say that these laws are also necessary. For an analysis of the modal status of these laws, see Section 7.5.
3. For example, Fales (1986 and 1990).
4. Freddoso (1986, 233–4) argues that it is "impossible that there should have been different laws of nature if by this we mean either (i) that the substances that, in fact, exist could have had different natures, i.e., could have instantiated different natural kinds, or (ii) that the natural kinds instantiated by these substances could have been necessarily tied to basic causal dispositions different from those they are, in fact, necessarily tied to." "Nonetheless," he says, "there could have been different laws of nature in the sense that it is metaphysically possible (i) that there should exist nat-

257

ural substances of kinds that have, in fact, never been instantiated and/or (ii) that the natural kinds that are, in fact, instantiated should not have been."

5. I am not of course the only one to seek to do this. Shoemaker (1980), for example, has done so. He holds that the totality of a property's causal potentialities constitutes its individual essence. (He calls this the "total cluster theory" in contrast with the view that the essence of a property is a proper subset of the cluster of causal potentialities.) This being so, he argues, "causal necessity is just a species of logical necessity" (p. 124). To the extent that causal laws can be viewed as propositions describing the causal potentialities of properties, it is impossible that the same properties should be governed by different causal laws in different possible worlds, for such propositions will be necessarily true when true at all (p. 124). I agree.

6. As Bigelow (1999) does in his reply on this point.

7. It is important to understand the difference between a global principle and a mere universal generalization. Where we have a global principle, the scope (see Section 7.3) of its antecedent or reference class is a whole ontological category. It is the class of all events and processes or all physical systems. Where we are dealing with a mere universal generalization, the antecedent or reference class is more restricted in scope.

8. Fales (1990, 218–9) says: "We may think of the universe of physical universals as being bound together into a structure. The binding relation, the cement which determines this structure, is the causal relation. Now our view (in its weaker form) is that it is part of the essence (what Armstrong calls the 'nature') of each universal that it is causally related to other universals in just the way that it is. Because the essences of the universals to which it is related are determined by the further universals to which they in turn bear relation R, and so on, it follows that the essence of each universal is in effect a function of the entire structure. It is an all-or-nothing affair: a change in the identity (essence) of one universal would ramify throughout the entire system. Hence the physical universals which are Rly connected to U – and this, directly or indirectly, includes all of them – are such that their essence would not be what it is if the relations in question did not obtain."

258

Part 5

The New Essentialism

8

The Essentialist Program

8.1 ESSENTIALIST AND HUMEAN METAPHYSICS

Scientific essentialism is a theory about natural necessity and also about the laws of nature. It is the thesis that there are natural necessities in the world that find their expression in cause and effect relationships and the laws of nature. If this claim is correct, then it follows that the laws of nature are immanent in the world, neither imposed on it by God, nor existing in it by chance. But what I wish to argue for is not just a theory about these things that is on a par with other theories about them. For scientific essentialism is also a metaphysic that has implications for philosophy right across the board – in ontology, epistemology, logic, and in most other areas. It is a thesis about the sources of power in the world, about the nature or reality, about the connections between things, about logical analysis, and even about the methodology of philosophical inquiry. As the dominant metaphysic, the Humeanism with which I have sought to contrast scientific essentialism also has broad implications in philosophy. It is also not just a theory of science, or language, or of what exists, but a metaphysic that gives shape to contemporary theories in all of these areas, and many others. Indeed, it is the dominant metaphysic of current Anglo-American philosophy.[1]

Some will say that the Humeanism I have described and criticized is unrepresentative of the philosophies of many of my contemporaries who would be happy enough to describe themselves as Humeans. But I make no apology for this. The version of Humeanism I have chosen to consider is one that is widely accepted.[2] I did not invent it, and it is not, as Bigelow (1999) suggests, "an extreme" form of Humeanism, but a fairly standard one. In any case, even if it were an extreme position, it would still be pertinent to consider it. For any theory that concedes that the identity of a thing might depend, even partly, on how it is disposed to behave would not be truly anti-essentialist, and therefore not one that stands in sharp contrast with the kind of theory I wish to present.

A metaphysic cannot be judged as a more specific philosophical theory might be. Because it is so wide-ranging, it has to be argued for in a different way. One cannot say, "Here are the problems; here is my solution," and then argue that this solution is better than any other. For one thing, the problems shift from one metaphysical position to another. Thus, for a Humean, there are two serious problems about laws of nature: the Necessity Problem (to explain the nature, or the illusion, of physical necessity),[3] and the Idealization Problem (to explain the existence of laws that appear to range over idealized systems – for example, closed and isolated ones). For an essentialist, these are not very serious problems. On the other hand, there are problems for essentialism arising from illusions of possibility, some of which have already been considered. Because of its basic role in shaping philosophical theories, a metaphysic has to be argued for, and defended, on many different fronts at the same time. The only way I know of doing this is to set it out carefully, displaying its range and overall coherence, contrast it with other positions, show how it deals with philosophical issues in various fields, and answer specific objections. The form of the argument is therefore, as Bigelow (1999) correctly notes, an argument by display. You show your wares and invite people to buy them. If your system strikes your readers as being simpler, more coherent, or more promising than any alternative for dealing with the recalcitrant difficulties of other systems, then this may be a good enough reason to buy it.

My aim in this chapter is to continue this process and to examine the implications of scientific essentialism as a metaphysic of nature. In Ellis (1999a),[4] my aims were somewhat more limited. I sought to show that there were serious problems with Humeanism (my ultimate target), which anyone defending a Humean theory of causality would have to deal with. I did not focus on any particular Humean regularity theory, such as Jack Smart's (1993) theory. For there is not much to be gained by showing that a particular Humean theory lacks credibility. What I sought to do was attack the underlying metaphysic.[5] Many others have shown, to my mind conclusively, that Humean regularity theories of causation and of causal laws are unsatisfactory. Rom Harré and Edward Madden (1975), Roy Bhaskar (1978), and David Armstrong (1983) all did so in their own ways. But their message is still largely unheeded, not because their arguments were not good, but because Humeanism is not just a theory about laws of nature or about causal connections. It is a deeply entrenched metaphysic, and the alternatives these writers proposed – an essentialist theory of causation and a theory of causal laws as relations between universals, could not easily be accommodated within this metaphysic.[6] Consequently, the message that the Humean theory of laws of nature is radically unsatisfactory still does not get through. Hume's theory of causation and of causal laws is seen as being in trouble, perhaps, but still, it is supposed, it must ultimately be defensible.

In this chapter, I will argue that scientific essentialism requires a different kind of thinking cap (to borrow a phrase from Herbert Butterfield) and a different program of analysis in philosophy – one that might aptly be called "realistic analysis" (in order to distinguish it from the kinds of semantic and "possible worlds" analyses standardly employed in philosophy). A different approach to philosophy is required by scientific essentialism because it changes the kinds of examples one can use and the kinds of modeling one can do in one's attempts to solve philosophical problems. The changes are required in order to accommodate the idea that the world consists, not of essentially passive things, but intrinsically active ones. One cannot, for example, think of a property as just a set of objects in some domain or other, as though the property had no powers, but was just a way of classifying the objects in this domain. If the causal powers of things are real properties, then things must be disposed to behave in certain ways in virtue of having these properties. The bearers of these properties must therefore stand in the relation of potential primary participant in a certain natural kind of process – namely, a causal process. Such facts as these must somehow gain recognition in our formal semantics and logical analyses.

8.2. THE ULTIMATE SOURCES OF POWER IN THE WORLD

As a thesis about the sources of power in the world, scientific essentialism stands in direct opposition to Humeanism. It is an opposition that echoes the eighteenth century conflict between the Newtonians on the one hand, representing the view that inanimate nature is essentially passive, and the Leibnizians on the other, who argued that things in the inanimate world have certain powers of action. According to mechanism, genuine causal powers are either non-existent (Hume) and therefore to be explained away (for example, as illusory) or they are derived from the agency of God (Newton, Berkeley). According to Leibniz, the things that exist in the world have powers, which he encouraged us to think of as causal powers. These are the powers he calls their living forces or *vires vivae*. While, strictly speaking, Leibniz did not believe in causal interactions – everything happens in Leibniz's theoretical world as a result of pre-established harmony – he nevertheless believed that the *vis viva* of a body is the true measure of its effectiveness in what we (naively) call "causal interactions."

In Leibniz's philosophy, *vis viva* (mv^2) was thought of as the force that animates things in nature. He thus distinguished his own conception of force sharply from Newton's concept of *"vis mortua."* For Newton, forces are actions externally impressed on bodies that are productive of changes of motion (changes of momentum). Hence, according to Newton, the correct measure of the force impressed on a body is just the change in momentum it undergoes – the quantity *mv*. Throughout the eighteenth century the question of which is the true measure of force – mv^2 or mv – was hotly disputed. This dispute was not, as d'Alembert says it was, just a dispute about words.[7] It was an absolutely fundamental disagreement about the sources of power in the world.[8] Do the laws of nature operate on an essentially passive world, as Descartes, Newton, and Malebranche believed, or are the things in the world animated by living forces, as Leibniz believed?

Historically, this argument between the Leibnizians and the Newtonians about the sources of power was not just philosophical byplay. It was a debate within the natural philosophical (scientific) community, engaged in by people who were themselves primarily natural philosophers, and it almost certainly had a major influence on the course of science itself. For Newton, forces are actions exerted on bodies that are productive of changes of motion ("momentum," in today's terminology). As actions, forces are always external to the bodies they act on, and are necessarily produced by objects, or beings, capable of exerting them (as explained in Section 3.3). But if inanimate brute matter is intrinsically inert, as most philosophers of the time, including Newton, believed, then the source of their motion must lie elsewhere. They have to be pushed or pulled around by agents of some kind. Forces can only measure the strength of this pushing or pulling.

For Leibniz, forces are effectively causal powers. They may be either inherent causal powers, like gravitation, or they may be causal powers acquired by motion. In either case, forces are clearly located in the material world, and have no dependence, direct or indirect, on agency. Leibniz argued (reasonably, it now seems to us) that it takes as much force (read "causal power") to raise a weight of one pound through four feet as it does to raise four weights, each of one pound, through one foot. Conversely, he argued that the force acquired by a weight of four pounds falling

264

through one foot must be the same as that acquired by a weight of one pound falling through four feet (as indeed it must be if force is *vis viva*). To us, this may seem both true and obvious. Nevertheless, the concept of *vis viva* was firmly rejected by Newton and by most natural philosophers in the eighteenth century. Indeed, rather than admitting the basic significance of Leibniz's concept of *vis viva,* some of Newton's followers went to what may now seem to be quite extraordinary lengths to avoid having to use it in explanations.[9] By the end of the eighteenth century, the Newtonians were considered to have won this debate, although the reasons why they did so have never been entirely clear. Whatever the reason, the triumph of the Newtonians over the Leibnizians in natural philosophy is probably one of the main reasons why neither the dynamical theory of heat nor the law of conservation of energy, for which the concept of *vis viva* is obviously central, were developed in the late eighteenth or early nineteenth centuries, when they clearly could have been.[10]

The triumph of Newtonianism in natural philosophy led naturally enough to the triumph of Humeanism in metaphysics. Like Newton, Hume denied that causal powers exist in inanimate matter. But, unlike Newton, he rejected the notion of causal power altogether, including divine and human agency. The laws of nature, he claimed, are just universal regularities. They are brute general facts – not divine commands. Nor are they the ways that things have to behave because of their natures. They are just universal propositions about the behavior of things in the world. If it seems to us that a cause necessarily produces its usual effect, then this appearance of necessity is just an illusion – a projection on to the events said to be causally related of our firm convictions or expectations concerning what sequences of events will occur. There are no necessary connections in nature,[11] Hume said, only the habits of thought established in those who observe it.

In this chapter, I will argue from an essentialist perspective, which, in spirit, if not in detail, is more in the tradition of Leibniz and Boscovich[12] than of Descartes, Newton, or Hume. Like Leibniz, I suppose that the laws of nature are grounded in intrinsically powerful properties. They are not imposed by God on things that are intrinsically powerless, nor are they just regular patterns of behavior that happen to be displayed by such things, just as if they were imposed by God.

8.3. FORCES, CAUSAL RELATIONS, AND CAUSAL POWERS

In his remarkable paper 'Remarks on the Forces of Inorganic Nature', J.R. Mayer (1842) argued for the principle that has since become known as the law of conservation of energy. This paper owes much to the revival of the Leibnizian tradition that occurred in Germany in the 1820s and '30s. Mayer argues that forces are one of just two species of causes – that is, things that are able to change form, but remain quantitatively the same throughout the process of doing so, so that the principle, causa aequat effectum, is satisfied. The other kind of cause is matter. For matter too can undergo transformations while remaining quantitatively the same. Forces are distinct from matter, however, in that they lack such qualities as weight and impenetrability. Forces, Mayer says, are *"indestructible, convertible imponderable objects."* But "so far as experience goes," he says, "[matter and force] never pass one into another." The conceptual unification of mass and energy is thus anticipated. Mayer goes on to argue that "falling force," or what we should now call "gravitational potential energy," and heat are both forces, and that any given quantity of either must therefore be equal to the quantity of *vis viva* into which it could in principle be transformed, or from which it could in principle be derived. He then uses some old data on the specific heats of gases at constant pressure and at constant volume to calculate how much heat is required to cause the expansion of the gas at constant pressure, and thus he is able to calculate, correctly in principle, the mechanical equivalent of heat.

It is inconceivable that this paper could have been written by a Newtonian. The conceptual framework required for it is completely at odds with Newtonianism. On the Newtonian conception of force, Mayer remarks:

Gravity being regarded as the cause of the falling of bodies, a gravitating force is spoken of, and so the notions of *property* and of *force* are confounded with each other: precisely that which is the essential attribute of every force – the union of indestructibility and convertibility – is wanting in every property: between a property and a force, between gravity and motion, it is therefore impossible to establish the equation required for a rightly conceived causal relation. If gravity be called a force, a cause is supposed which produces effects without itself diminishing, and incorrect conceptions of the causal connexion of things are thereby fostered. In order that a body may fall, it is no less necessary that it should be lifted up, than that it should be heavy or possess gravity; the fall of bodies ought not therefore be ascribed to their gravity alone (p. 199).

Most philosophers nowadays would, quite properly, distinguish between the Newtonian forces acting on a body in a given situation and the causes of its resultant motion. But Mayer is right about one thing: Causal processes are energy-conserving transformations of states of affairs involving several objects (or the several parts of objects). But forces of the Newtonian kind nevertheless have a vital role in the analysis of such transformations. They tell us about the contributions of the various causal powers of the objects involved in them to what occurs so that once we know what the causal powers are, how the objects involved are placed with respect to each other, and how they are disposed to respond to the actions of these powers, then we can predict the ways in which they will behave, and with what probabilities.

While objects are the bearers of the causal powers, and of the capacities to respond to them as they do, it is evident that the causal powers of things stretch out somehow beyond their boundaries. If the world were a Newtonian world of atoms in empty space, it would be mysterious how one object could affect another without mediation. According to Newton, whose authority on questions of natural philosophy dominated the eighteenth century, all causal power derives from God. It does not reside in matter, as many later Newtonians came to believe. So, if one object acts on, or is attracted by, another, this is not because of the intrinsic natures of these objects. Ultimately, it is because this is what God commands. In a letter to Mr. Bentley in February 1692, Newton said:

It is inconceivable that inanimate brute matter should, without the mediation of something else which is not material, operate on and affect other matter without mutual contact, as it must be if gravitation, in the sense of Epicurus, be essential and inherent in it. And this is one reason why I desired you would not ascribe innate gravity to me. That gravity should be innate, inherent, and essential to matter, so that one body may act upon another at a distance through a *vacuum*, without the mediation of anything else, by which their action and force may be conveyed from one to another, is to me so great an absurdity that I believe that no man who has in philosophical matters a competent faculty of thinking can ever fall into it. Gravity must be caused by an agent acting constantly according to certain laws, but whether this agent be material or immaterial I have left to the consideration of my readers (Thayer, 54).

My answer to the problem of action-at-a-distance is not to deny the reality of the causal powers, but to deny the absoluteness of the visual and tactile boundaries of things. Things must somehow stretch out beyond their boundaries to be involved in causal processes in the ways in which they are. The spatio-temporal distortion that we say is "produced" by the presence of a heavy object is not, I should say, an effect of its presence but an integral and essential part of it, without which the object could not exist. The distortion of the surrounding electromagnetic field within which an electron is (roughly) located is likewise not something that exists independently of the electron's existence. Plausibly electrons just are discontinuities of some kind in the electromagnetic fields that we say they generate.

Newtonianism is no longer a powerful force in natural philosophy, although there is no obvious successor to it. The view of reality that science now seems to be forcing us to take has nothing of the neatness of self-contained atoms in Newtonian space and time. Rather, it is a confused picture in which space and time are just aspects of the complex space-time-energy structure of General Relativity, and atoms are wave-packets of some kind, which are anything but self-contained. But Humeanism remains a powerful force in philosophy despite the fact that it no longer has a plausible basis in natural philosophy as it once did. It is time perhaps to consider seriously what a contemporary metaphysic should look like. For the remainder of this chapter, I will discuss some of the ways in which scientific essentialism provides new ways of looking at old problems and suggests new programs of philosophical analysis. In what follows, I will consider the limitations of the program of logical analysis and suggest a more realistic program, the possible worlds theory of counterfactual conditionals, and propose an analysis that does not require belief in any worlds other than this one, the problem of induction, which is largely a problem generated by Humeanism, and the problem of singular causation, which is certainly a problem for Humeans, but not a problem for essentialists.

8.4. THE PROGRAM OF LOGICAL ANALYSIS

The belief that nature is intrinsically powerless and that the laws of nature are brute general facts about how the intrinsically powerless things in the world are or behave leads naturally enough to certain ways of representing the world in our theorizing. For example, it encourages us to think of the world as a totality of self-contained logically independent facts, or states of affairs, more or less as Russell did, since what exists at any one place or time, we are led to believe, must be independent of what exists at any other place or time. Given this conception of reality, it is hard to think of the laws of nature as anything other than universal regularities that happen to exist within this totality of facts.

In general, our descriptions of the world will be true iff they correspond to the facts. Given a Humean conception of reality, these facts must be non-modal, as Bigelow and Pargetter (1990) insist – that is, they cannot be such as to imply the existence of any necessary connections between distinct states of affairs. The true descriptions of the world can therefore make no reference to causal powers or other modal properties, unless the existence of these properties can somehow be reduced to non-modal ones. The primary problem for the theory of truth, therefore, is seen as being to develop a non-modal language that is adequate for describing the world, and a correspondence theory of truth that is adequate to explain what makes any true statement expressible in this language true. The theory of truth-preservation, and therefore the theory required as a foundation for logic, must likewise depend on the development of such languages and appropriate theories of truth for them. A Humean conception of reality thus lies behind, and motivates, the development of extensional logics with extensional semantics, and underwrites the deployment of such languages for describing the world.

Given this conception of reality, the laws of nature are naturally enough supposed to relate classes of things rather than describe their causal powers. The law that all As are Bs, for example, becomes simply the fact that the class of As is included in the class of Bs. There is no suggestion that there might be, let alone must be, a reason why As are Bs, if in fact it is a law that all As are Bs. Nor is there any suggestion that the required reason must have something to do with what it is for something to be an A. Indeed, the idea that there might be some connection between a thing's being an A and its being a B, which is responsible for its being a law that all As are Bs, just drops out of the picture.

In many cases, it is true that truth seems to be simply a relationship of correspondence with non-modal facts. If something is referred to by name, and something is then said about it, then what is said is true if and only if what is referred to is the way it is said to be. This seems obvious. However, there are many different kinds of propositions that are held to be true or false, including many whose relationship to reality is quite obscure. The program of trying to specify realistic truth conditions for these more troublesome propositions, including statements of laws and causal connections, has been a preoccupation of philosophers for most of the twentieth century. One way of dealing with them was that pioneered by Russell, Moore, and others. These philosophers sought to analyze the troublesome propositions logically, so that their truth or falsity could be derived from that of more elementary propositions whose truth or falsity conditions were not thought to be problematic, or, at any rate, less problematic than those that were to be analyzed. This was the program known as "logical analysis."

The program of analysis has often been pursued subject to two important constraints. The first of these is the *requirement of extensionality,* which is dictated by Humeanism, and is a constraint on the kind of analysis that is acceptable. To explain what makes a given proposition true, it is supposed that we must be able to express this proposition in an extensional language – that is, a language whose terms refer to things in a specific domain and whose predicates refer to sets of things in this domain. It is not a language in which we can refer to the kinds or properties of things that exist in this domain. Languages, like natural languages, in which reference can be made to properties and kinds are not extensional languages.

The second common constraint on the adequacy of any proposed analysis is the *requirement of realism.* The domain in which the language is interpreted must be a domain of real things. One cannot accept as satisfactory an analysis that refers to a domain of things one does not believe in. For the aim of the program is to explain the manner in which true propositions correspond to reality. It is to specify the truth-makers for the propositions we believe to be true, and explain the nature of the relationship of correspondence between the truth-makers and the propositions they make true. It is not good enough, for example, to say that

$$2 + 2 = 4 \text{ is true iff } x + x = y \text{ is satisfied by the sequence } \langle 2, 2, 4 \rangle$$

if one does not believe in numbers. The program aims to provide realistically acceptable truth conditions, in which reference is made to real existents. If any otherwise satisfactory analysis failed to meet this requirement, then an analysis of the analysandum would be needed.

The program of analysis has often been pursued without much regard for the requirement of realism. "Possible worlds" analyses of modal and conditional propositions would seem to violate this requirement. Of course, there are some philosophers who think that these analyses do satisfy the requirement of realism. They believe in possible worlds. But there are many more who do not think this, and who, accordingly, have sought to analyze propositions about possible worlds so that they can be realistically understood. There are also some purists who consider the question of the reality of possible worlds to be relatively unimportant. What interests them is what can be done with this kind of analysis by ringing the changes on accessibility and similarity relationships between possible worlds. For my purposes, though, the requirement of realism cannot be ignored. For, unless we can provide realistic truth conditions for the propositions we believe to be true, we should be unable to explain, in correspondence terms, what makes them true. So we cannot ignore the requirement of realism compatibly with the basic aim of the program. The issue is a major one for the analysis of modal and conditional propositions because the analyses that are formally the best are ontologically the least acceptable. Any alternative account should therefore be able to explain why the possible worlds theory works as well as it does.

Ontologically, the main trouble with possible worlds (other than this one) is that the only reason anyone has, or ever could have, to believe in them is that they are needed, apparently, to provide truth conditions for modals and conditionals. They are needed, it seems, because if this world is the only reality, and reality is non-modal, then there is not enough reality to go around. The truths expressed in modal and conditional propositions are left with nothing in the Humean non-modal actual world to correspond to. To explain what makes them true, it seems, we should need things that could or might exist, as well as the things that do exist. Therefore, it is argued, this world cannot be all there is. There must be other possible worlds too – worlds that are not actual, merely possible. This is why some people believe in the existence of merely possible worlds. It is indeed the only legitimate reason anyone could possibly have for believing in the existence of any such world. There could not be a good reason of another kind for believing in such a world because, if there were such a reason, the world in question would not be merely possible. Suppose, for example, that according to some accepted symmetry principle, there must exist a kind of mirror image of this world from which we have been causally isolated since the Big Bang. If we really believed in this symmetry principle, then of course we should not regard this other world as being merely possible. We should simply think of it as a remote and inaccessible branch of our own world.

The argument that belief in the existence of merely possible worlds cannot be independently justified looks like a devastating argument against any literal interpretation of possible worlds theory. Moreover, if the correspondence theory of truth requires such an interpretation, then this is a devastating argument against the correspondence theory. If you have to invent a non-denumerable infinity of possible worlds for which you could not possibly have any other evidence, just to save your theory, then your theory cannot be much good. This would be ad-hocism with a vengeance.[13]

But maybe one does not have to be a realist about possible worlds to save the correspondence theory of truth. The failure of the program of analysis to discover realistic truth conditions may be due to something else.

It may be due to the inherent Humeanism of the program. Indeed, the program is not only Humean, it is also nominalistic, and does not even recognize the existence of genuine properties and relations. For one does not have properties or relations in this tradition, one only has predicates. One does not have names denoting kinds of things, such as copper or tigers, one has the predicates "is copper" or "is a tiger." Properties, relations, and substances, if they can be talked about at all, are identified with their extensions. The property of being yellow is the set of all yellow things. The relation of being *greater in mass than* is the set of ordered pairs x and y such that x is greater in mass than y. The substance, copper, is the set of all things that are coppery. It is as if the world consisted just of individuals and sets of individuals, as indeed Quine, Smart, and many other Humeans evidently believe. However, the world certainly consists of things belonging to other ontological categories, and to describe it adequately we need a language in which we can talk about and relate these different kinds of things. The failure of the program of analysis may therefore be due simply to the poverty of the semantic theory on which it depends, and ultimately to the poverty of Humeanism.

If the ontology of the real world is as rich as that described in Chapter 2, the problem of specifying realistic truth conditions for modals and conditionals is unlikely to be overcome unless we start with a much richer base than the lean ontology of things and sets of things accepted by most logical analysists. To deal with this problem, we should begin to construct our program of analysis on the basis of an ontology that also includes natural kinds of objects and processes, and dispositional properties of various kinds, so that natural modalities may be explained with reference to things existing in the actual world. The program should not be one that seeks to reduce everything at the outset to individual things and sets of things.

8.5. ACTUAL AND POSSIBLE WORLDS

According to possible worlds realists, the actual world is just one of infinitely many possible worlds—namely, the one we happen to inhabit. Otherwise, there is nothing special about it. A possible world is real, but not actual, if it is not the one we live in. The phrase "actual world" is therefore to be understood as a token reflexive, just like the words "now" and "here," but one that serves to locate ourselves in the realm of possibilities rather than in time or space.

I have no strong objection to using the word "actual" in this way. However, since I do not believe that we could possibly inhabit any world that is not of the same natural kind as ours, we should have to distinguish two classes of possible worlds – those that could be the actual world and those that could not be. However, I would prefer to use the word "actual" to denote what exists as a matter of fact, whether or not it is causally accessible to us. To refer to the world we inhabit, I shall simply say "this world." Logically, "this world" behaves like the phrase "this island" to a person who does not know whether there are any other islands, and is necessarily incapable of finding out. A person using this phrase can presumably conceive of the possibility of there being other islands, but can have no sufficient reason to believe that there are any.

The other worlds that possible worlds realists talk about are supposed to be real worlds, indeed, no less real than our own. But, according to possible worlds realism, all such worlds exist necessarily. That is, their existence is not contingent. This stands in stark contrast to the position I wish to defend. To believe in the possibility of other worlds than this one, one does not need to believe that every possible world exists, let alone that it exists necessarily. On the contrary, I should think that most of the possible worlds which we are able to imagine probably do not exist. Nor does the essentialist theory of natural necessity developed in the last chapter require belief in the existence of all these possible worlds. It does not even require belief in a multiplicity of worlds. There just might not be any other worlds.

A possible worlds realist is like someone who believes that the ocean is an infinite expanse, and that every conceivable island exists somewhere on this vast ocean. I see no good reason to believe in any such fantasy. To pursue the analogy, it is plausible to suppose that there are some other islands around somewhere, because it would be surprising if ours were the only one, or if our kind of island were the only kind. But that is about all we can say. It is not an *a priori* truth that other islands exist, and direct evidence of what exists beyond the range of possible causal influence is impossible. If the world is one of a natural kind, as I suppose, and the most general laws of nature simply describe the essential structure of worlds of this kind, then these most general laws would be similar in nature to the more specific causal laws we have already discussed. The global laws would be true in virtue of the world's essence, just as the more specific laws are true in virtue of the essential natures of the things of the more specific kinds that can occur in worlds like ours.

If this is right, and the global laws of nature do derive from the essential nature of the world, then this provides a very good explanation for the kind of necessity that these laws manifestly do have. They are not *a priori,* since it is *a posteriori* what kind of world this is, and what kinds of things are the constituents of such a world. That is, these things have to be discovered by empirical investigation, and it could turn out that we do not live in the kind of world we think we live in, or that it contains things of natural kinds different from those we suppose it to contain. On the other hand, if the laws of nature derive from the essential properties of worlds of this kind, and of things of the kinds that worlds of this kind necessarily contain, then we know that these laws cannot be violated in any world of this kind.

Evidently, then, we are not forced to silence on what would have to be the case in any world of the same natural kind as ours. For we know that any such world would have to behave according to the same natural laws, and this is something we can say whether or not we believe there are any other worlds. Thus, it begins to look as though we may be able to do something like possible worlds semantics without postulating the existence of any other possible worlds. But why bother to do this? What is wrong with current possible worlds semantics? One thing that is wrong is that the ontology of possible worlds realism is unacceptable. But, more fundamentally, what is wrong is the inherent nominalism of the whole program of analysis that gives rise to possible worlds semantics.

8.6. REALISTIC SEMANTICS FOR NATURAL MODALITIES

If the world is one of a kind, then any world of the same kind must be similar to ours in certain ways. Specifically, it must, qua world of this kind, have the same essential properties and structure. The essential properties of our world include its having the same basic ontology, since our world is essentially a world of a kind in which the things that do exist can exist. So any other world of the same kind must also have this basic ontology. Worlds of the same natural kind must also have the same space-time structure, evolve in the same sort of way (for example, from a Big Bang, if this is how our world began), display the same global symmetries, and be restricted by the same conservation laws. Let us say that any such world is one that is *similar* to ours. Then, since we know from our own case quite a lot about the essential properties, constitution, and structure of the natural kind instantiated by our own world, we can say what must be the case in any world of this kind, including our own.

We can now use the idea of worlds similar to our own to give an account of the truth conditions for natural necessity and possibility claims – an account that is something like a possible worlds account, but that is nevertheless importantly different from it. We can provide this new account without assuming the existence of any world other than this one. For once we know what the essential nature of this world is, we know what must be the case in this, or in any other world of the same kind, and hence in any similar world, if any other such world should happen to exist. We know what must be the case because any instance of a natural kind must have the essential properties and structure of that kind and be similarly constituted. Consequently, any instance of a world of the same natural kind as ours must have the same global properties and structure and be constituted by objects, properties, and processes of the same fundamental natural kinds.

Given this concept of a world similar to our own, a naturally necessary proposition may be defined as any proposition that must be true in any world similar to this world. A necessarily false proposition, in the sense of natural necessity, may accordingly be defined as any proposition that must be false in any similar world, and a naturally contingent proposition as any that could be either true or false in a world similar to our own. The necessarily true propositions will thus be those that hold in virtue of the global properties, structure, or constitution of worlds similar to our own; the necessarily false ones will be those that are incompatible with these properties, or with this constitution or structure; and the contingent propositions will be those that are compatible with, but not necessitated by, these properties or this constitution or structure.

The most general laws of nature, the global laws, understood as propositions about this world, are all naturally necessary by this definition. For these laws define the global properties and structure of worlds that are of the same natural kind as ours. Consequently, they must hold in all worlds similar to ours.

Similar worlds must also have the same basic ontology of kinds of objects, properties, and processes. It must, for example, be a physical world made up of particles and fields of the same fundamental natural kinds as those that are fundamental in this world. If electrons and protons are such fundamental natural kinds in this world, then they must also exist in every similar world. If mass and charge are fundamental natural kinds of properties, then mass and charge must exist as fundamental kinds of properties in every world that is similar to ours. If electromagnetic radiation is a fundamental natural kind of process, then electromagnetic radiation must occur in every world of the same natural kind as ours. If, in a given world, any of these kinds of objects, properties, or processes did not exist, then it would not be a world of the same natural kind as ours. If we accept this analysis, then we are committed to saying that some natural kinds of objects, properties, and processes exist or occur necessarily. The propositions that are naturally necessary will therefore include some open existential statements – propositions that Popper would have said were metaphysical, and therefore do not belong to the body of science. This is perhaps a surprising result. It is one that as far as I know, distinguishes the present theory of natural necessity from every other such theory. Nevertheless, it is not implausible to say that whatever exists most fundamentally in this world in any given category exists necessarily. For whatever exists most fundamentally in a given category cannot be ontologically dependent on the existence of anything else in that category. Therefore, if it is plausible to speculate that such things as electrons and protons and electromagnetic radiation exist fundamentally in this world, then it is also plausible to speculate that they exist necessarily.

It is not, of course, being said that there are instances of any of these kinds of objects or processes that exist necessarily. It is not that this or that electron or this or that instance of light emission exists necessarily. What is necessary is only that there be instances of these kinds in this kind of world. However, given that there are instances of these kinds, it follows that there are some specific instances of some basic quantitative properties that must be instantiated. For if electrons exist, then unit mass and unit charge must also exist. So not only do mass and charge exist necessarily in this and similar worlds, so also do unit mass and unit charge. Could there be fundamental natural kinds of objects, properties, or processes existing in worlds similar to ours that do not exist in our world? In other words, could a world of the same natural kind as ours have a richer basic ontology? I think not. A world with an ontology otherwise like ours, which included some extra ingredients, could be a species of world of the same genus as ours, but it could not be a world of the same specific natural kind as ours. Things of the same specific natural kind must be essentially the same. But if the essence of A is included in the essence of B, then B can only be a species of A. Worlds with different basic ontologies cannot be essentially the same.

Remember that I am talking here about basic ontologies, not derivative ones. I am not saying that every natural kind (whether basic or not) that exists in our world must exist in every similar world, or conversely. For some of the natural kinds that exist in this world and that are not ontologically fundamental may not be able to exist in another world of the same kind. The conditions that happen to obtain in this other world might not favor them. Conversely, there could be natural kinds that are not fundamental, exist in other worlds similar to ours, but do not exist in our world. It is also important to note that natural necessity is not being defined here just in terms of truth in similar worlds. A naturally necessary proposition would of course have to be true in every similar world, but its being true in every such world is not what makes it necessary. It is not even being assumed that there are any other worlds, whether similar to our own or not. There may or may not be any. On the account being proposed here, natural necessity derives from the essential natures of things – of worlds and their contents.

Natural possibility for a given world is just consistency with the essential nature of the kind of world it is, and with the essences of the various kinds of things it necessarily contains. If it is possible that *p*, in this sense, and also possible that not-*p*, and the world happens to be a *p*-world, then this is an accidental feature of the world. It is something that might or might not have been the case. Hence, there could be a world of the same natural kind as ours in which '*p*' is true. This possibility is not ruled out by the nature of such a world or any of its constituents. I do not say that there is such a world, given that "It is possible *p*" is true. That is David Lewis, not I. But I do say that there could be such a world, given the nature of the world in which we live.

I wish to stress that the modalities of natural necessity and possibility defined here are not just *de dicto* modalities. For they are grounded in the essential natures of things. Natural necessities and possibilities are properties of the real world that are discoverable by the ordinary procedures of scientific investigation. Indeed, if I am right about the basic aim of research in the natural sciences, it is precisely to discover what is possible or impossible in this world, and hence in any other worlds similar to ours. Thus, as I understand them, natural necessities and possibilities are *de re* modalities, not *de dicto* ones.

It may be objected that natural necessity, as here defined, is not necessity in the strongest metaphysical sense. Metaphysical necessities are propositions that are true in all possible worlds (although not under all interpretations of their non-logical terms, as formal logical necessities are). Propositions such as "electromagnetic radiation is propagated according to Maxwell's equations" are plausibly necessarily true in this metaphysical sense, if they are true at all. But a proposition such as "electrons exist" is not so plausibly metaphysically necessary. For it is not implausible to suppose that there could be a world, albeit not one of the same natural kind as ours, in which electrons do not exist. But the proposition to be considered is not "electrons exist," but rather, "electrons exist in our kind of world," – that is, in the actual world you and I inhabit (and in any other world of the same natural kind as this one). This proposition is certainly true. It is also necessarily true in the required metaphysical sense, if any world of the same natural kind as ours must be one in which electrons exist.

The same is true of the global laws we have been discussing. The proposition that every event or process is intrinsically conservative of energy does not sound like a necessary proposition. But it does if you add that the events or processes within your scope of predication are those occurring in this world, or in any other world of the same natural kind as ours. So understood, I claim, this proposition is metaphysically necessary. There could not be a world of the same natural kind as ours in which energy is not so conserved.

8.7. AN ESSENTIALIST THEORY OF CONDITIONALS

Given these concepts of natural necessity and possibility, it is not difficult to see how one could develop a corresponding theory of conditionals. Such a theory would be just like a possible worlds theory, except that it would refer only to what must, might, or could not happen in worlds similar to ours (if indeed there are any such worlds). It would not refer to what does happen in some or all possible worlds that are accessible to ours, as the standard theory does. To evaluate a conditional, on such a theory, we should have to consider what would happen, or be likely to happen, in a world of the same natural kind as ours in which the antecedent condition is satisfied, other things being as near as possible to the way they actually are. The proposition "if A were the case, then B would be the case" will be true on such a theory if and only if in any world of the same natural kind as ours in which "A" is true, in circumstances as near as possible to those that actually obtain, "B" must also be true. It would not be a question of whether "B" is true in every such world, for there might not be any such world, or it might happen by accident that in every world of the same natural kind as ours in which "A" is true, "B" is also true. The truth of the conditional would not depend on what other worlds, if any, there may happen to be. What would make the conditional true is what must obtain in any world of the same natural kind as ours in which "A" happens to be true and in which the appropriate conditions of similarity of circumstances are satisfied.

Truth conditions for conditionals may thus be specified without assuming that there are any worlds other than this one. There are of course difficulties in defining the antecedent conditions for any given conditional. For it is not always clear how to define "circumstances which are as near as possible to those which actually obtain, but in which 'A' is true." But this is a difficulty which a realistic theory of truth conditions for conditionals would have in common with possible worlds theories. And it is as well that this should be so. For a theory of conditionals that removed this inherent vagueness would immediately, and properly, be suspect.

8.8. REALISTIC ANALYSIS AND THE THEORY
OF CONDITIONALS

In his paper "Scientific Ellisianism," Bigelow (1999) argues that there are serious difficulties with this essentialist theory of truth conditions for conditionals. Specifically, far too many conditionals are likely on careful analysis to have impossible antecedents, and therefore turn out to be vacuously true. For although it may be easy enough for us to imagine a world very like ours in which a false antecedent is true, it may well be really impossible that there should be such a world. I am persuaded that this objection is indeed a serious one. At the same time, the resolution of this difficulty has important implications for the methodology of philosophical inquiry. Essentialists cannot conduct philosophical investigations in the same debonair fashion as Humeans, allowing their imaginations free rein in the construction of metaphysically possible worlds. They have to be much more aware of the limitations of the imagination as an instrument for determining what is really a possible state of affairs. For truth conditions for conditionals, the requirement of realism makes heavy demands.

As a scientific essentialist, I have to agree with Bigelow that our intuitive judgments of real possibility are fairly unreliable. Therefore I cannot be sure that any situation I care to imagine that would make the antecedent of a given counterfactual conditional true is a real possibility. I can perhaps be sure that if it were true, then the consequent would have to be true if the world is one of the same natural kind as ours. However, the world I have imagined might not be a world that is really possible. Therefore, the counterfactual conditional might really be only vacuously true.

As philosophers, we have been systematically trained to think that whatever is imaginable is possible. This is part of our Humean upbringing. Because we have been trained in this way, it is very easy for us to confuse epistemic possibility with real possibility. If it is imaginable, we suppose, then it must be at least logically possible. Take almost any philosophical paper written in the Anglo-American tradition. In it you will find examples of allegedly possible states of affairs. These supposedly possible states of affairs, which are drawn from the imagination, are the examples and counter-examples used by philosophers in our tradition in our arguments for the philosophical positions we hold. But our imaginations work without scientific constraint, and hence without having to consider the sorts of limitations on possibility that exist in the real world. Consequently, the examples used by philosophers often seem absurd to scientists and others who are not well trained in philosophy, and it has to be explained patiently that as philosophers, we are concerned, not with *physical* possibility, but only with *logical* possibility, the test for which, apart from there being no obvious contradiction in describing the case, is just imaginability or conceivability. At any rate, what is possible is supposed to have nothing to do with what is possible from a scientific point of view.

The methodology of testing philosophical positions by using imaginatively constructed, but scientifically implausible, or even physically impossible, counter-examples is a reasonable one if you are a Humean and believe that all events are loose and separate. For, in a Humean world, anything that is consistently imaginable is possible because the laws of nature in such a world can easily be changed to make it conform to the way we imagine it to be. The identities and natures of things never get in the way. But in the kind of world that science has revealed to us, we can no longer be so sanguine about the looseness or separateness of things, and our imaginations, which are necessarily superficial and cartoon-like in their representations, can no longer be considered reliable sources of information about real possibilities. A scientific essentialist has to be much more cautious than a Humean when making judgments about what is really possible. A scientific essentialist cannot, for example, rely on arguments from ignorance to establish that anything is really possible. For the fact that something is not known to be impossible does not imply that it is possible in the required metaphysical sense. It only shows that it is "epistemically" possible (that is, "possible for all we know"). But epistemic possibility is a different concept from real possibility, even though it is related to it. Nevertheless, epistemic possibility is a concept that is widely used, and it is undoubtedly one that is important in connection with the theory of conditionals.

It is in connection with the theory of conditionals that Bigelow sees trouble for scientific essentialism. He is not worried that some counterfactual conditionals with antecedents that are obviously physically impossible – such as "if we could walk on water, then we should have less need for boats" – should turn out to be vacuous. What concerns him is that many quite ordinary-looking counter-factuals such as "if I had a cold beer in front of me, then I should drink it" might well turn out to be vacuous. For, on Lewis's theory of counterfactuals (Lewis, 1979), the truth of this proposition requires that there be a possible world with the same laws, and with a past that is at least very like ours, except that miraculously, there is now a cold beer in front of my counterpart in this world. The miracle required might only be a small one, as he says, but it is a miracle nevertheless, and its occurrence might require that some causal laws be violated in a possible world that is retrospectively otherwise much like ours.

There are two ways of dealing with this objection. The first is to argue that the world's future, and the futures of all worlds of the same natural kind as ours, are at all times fairly wide open. This is plausible because, as we have seen, worlds of our kind all evolve by a process akin to a random walk. Consequently, almost any state of the world we may care to imagine that is not in itself incompatible with the laws of nature is one that might really exist in some world of the same natural kind as ours, and many of these states of affairs might indeed really have existed in our own world. We do not have to back track all the way to the Big Bang to realize the conditions envisaged when, for example, I say "If there were a beer in front of me, I should drink it." Given the openness of the future in all worlds of the same kind as ours, the state of affairs envisaged is obviously a real possibility. There are billions of ways in which it really could have come about.

The second way of dealing with the objection is to concede that many counter-factuals, and perhaps some indicative conditionals, might turn out to be vacuous, but to argue that what matters for conditionals is not whether they are true but whether they are assertible. For example, we might argue that conditionals are assertible if the information they provide indirectly about the world, in particular about the dispositional properties of things in more or less realistic circumstances, is both true and appropriate.

There are arguments for and against these two strategies. On the one hand, it is clear that we live in a highly indeterministic world, and that the future in any world like ours is, and always has been, open. Consequently, there must be a huge, indeed, non-denumerably infinite, range of really possible states of affairs to consider, when we wish to decide whether a given counter-factual conditional is true. There are plausibly very many ways in which it might really have come about that there is currently a beer in front of me, or in front of someone very like me, in my sorts of circumstances, and with a thirst, very like mine, in this world, or in another world of the same kind as this one. However, this is all a bit speculative. Many assertible counter-factual conditionals will no doubt slip through Bigelow's net in this way, and have to be counted as substantively true. But some may not, and we cannot know which. However, a counter-factual conditional would be no less assertible even if it were just vacuously true. Consequently, I think we must in the end just have to settle for assertability conditions for conditionals, and forget about their truth conditions.[14] Their substantive truth or vacuousness is simply not relevant to their assertability.

The attempt to provide adequate real possible world semantics for conditionals fails because the evaluation of conditionals does not require us to consider what might exist in other really possible worlds. A knowledge of the dispositional properties of things, of the actual circumstances, and a good imagination, are all that are needed. The causal conditional "$C_i x,y \Rightarrow E_i x,y$" is assertible, for example, iff

1. In the imagined circumstances, arrived at by making whatever changes to the actual circumstances are minimally required to accommodate the antecedent supposition $C_i x,y$, a dispositional property triggered in these circumstances, and having the consequent $E_i x,y$ as its outcome, exists.
2. There are no countervailing dispositional properties possessed by any of the individuals in the imagined circumstances, and also displayed in these circumstances, that are strong enough to overcome or swamp the display of the dispositional property having the outcome $E_i x,y$.

But what about truth conditions? We can easily imagine things to be the case that could not possibly be so. Therefore, we cannot just assume that the assertability conditions for causal conditionals given here are also truth conditions. The truths relating to causal conditionals are the underlying ones on which their assertability depends. That I am thirsty, for example, is a fact about me. It is also a fact that I like beer, and that there is nothing in the world that I would like better at the moment. Consequently, I would say, "If there were a beer in front of me I should drink it." It simply does not matter whether it is really possible for there to be a beer in front of me. And if this makes it vacuous, then it does not matter whether it is vacuous. By asserting the conditional, I tell you graphically what my desires are at the present time, and what you could do to satisfy them. It is better than saying "I am thirsty," because you might then offer me water, which is not what I want most. It is better than saying "I am thirsty, and I like beer," because this is compatible with my not wanting a beer at the moment.

For me, assertability conditions, or acceptability conditions as I prefer to call them, are enough. As I have argued elsewhere (Ellis, 1979), they are all we need for the evaluation of arguments involving conditionals. Indeed, it was primarily because of the implausibility and the unrealism of possible worlds semantics for conditionals that I sought to found logic, not on truth conditions, but on acceptability conditions. An argument is valid, I argued, iff there is no rational belief system in which its premises are accepted and its conclusion is rejected. And, given such a concept of validity, it is obvious that acceptability conditions are all that are needed for the purposes of logic. Now I find myself forced to the conclusion that acceptability conditions are more important for a different reason. Realistic truth conditions for conditionals can be defined fairly easily, given a semantics of real possibilities. But they are very hard to assess, and are of doubtful relevance to everyday discourse. Scientific essentialism thus provides a more sophisticated motivation for abandoning truth semantics, and moving over to an acceptability semantics of the kind described in Ellis (1979).

8.9. ESSENTIALISM AND THE PROBLEM OF INDUCTION

If scientific essentialism motivates reconceptualizing logic as the theory of rational belief systems and abandoning the Fregean conception of logic as the theory of truth preservation, as I think it does, it also has profound implications for the theory of rationality generally. In particular, it promises to transform our thinking about scientific rationality and the theory of inductive reasoning. If one believes, as Hume did, that all events are loose and separate, then the problem of induction is probably insoluble. Anything could happen. But if one thinks, as scientific essentialists do, that the laws of nature are immanent in the world, and depend on the essential natures of things, then there are strong constraints on what could possibly happen. Given these constraints, the problem of induction may well be soluble. For these constraints greatly strengthen the case for conceptual and theoretical conservatism, and rule out Goodmanesque inferences based on alternative descriptions of the world. This may not in itself solve the problem, but it significantly changes its nature.

The identities of things in Hume's world are independent of what they are disposed to do – that is, independent of their causal powers, capacities, and propensities. The dispositions of things are supposed to depend on the laws of nature, which might be different in different worlds. Thus, things that are constituted very differently might be disposed to behave in exactly the same way, if the laws of nature were sufficiently different (cf. the Catholic doctrine of transubstantiation), while things that have precisely the same constitutions might not, or might not always, be disposed to behave in the same ways in the same circumstances. In Hume's philosophy, the ways in which things are disposed to behave are supposed to depend, not on their intrinsic natures or constitutions, but on what the laws of nature happen to be. Therefore, for a neo-Humean, there is no solution to Hume's problem to be found by considering what sorts of things exist in the world.

Nor is Hume's problem solvable within a Cartesian, Lockean, Newtonian, Berkeleyan, or Kantian theory of reality. For on all of these theories, the world is essentially passive, and the ways in which things are disposed to interact are contingent on the laws of nature. Things, as they are in themselves, are supposed to have no genuine causal powers. If they appear to be causally interactive, then this is not due to any powers they may have by nature, or may have acquired, but to how they are required to behave by the laws that govern them. Metaphysically, things in the world are to be thought of as puppets pushed around by the forces of God or nature. They are not themselves actors on the stage. Induction is therefore a problem for the broad philosophical tradition that has its roots in seventeenth and eighteenth century mechanism. It is not just a problem of empiricism. It is a problem for anyone who believes that the laws of nature are superimposed on a world that is essentially passive, and that these laws are contingent, and not knowable *a priori*.

To solve the problem of induction, it may be necessary to break with this whole way of looking at things. What may be required is a conception of reality that denies that nature is essentially passive, or one that accepts that events may be necessarily connected. For Hume's thesis that all events are "loose and separate," and consequently that there are no necessary connections between events, is an ontological claim that if true, would seem to make the problem of induction insoluble. Of course, a solution to the problem of induction must ultimately be epistemological. That is, it must vindicate most of our ordinary scientific inductive practices. But how we ought to reason about the world might well depend on what kind of world we think it is. For this will affect what we think the epistemic task is. If we think the world is a Humean kind of world – that is, a world of disconnected events – then maybe no inductive strategies can be vindicated. If so, then perhaps this is sufficient reason to believe that it is not after all a Humean kind of world. If, however, we think it is a highly structured and necessarily connected world of natural kinds of objects, properties, and processes, as scientific essentialists do, then our aim must be to discover its nature and structure. And this is a very different task from that which confronts the Humean.

From the perspective of a scientific essentialist, all scientific inference is seen as depending ultimately, not just on observed regularities, but on what postulates about natural kinds are justifiable. The role of postulates about natural kinds in scientific reasoning has been stressed by a number of writers in the last twenty-five years (Butts, 1977, Forster, 1988, Macnamara, 1991). More recently, the point was made by Howard Sankey in his "Induction and Natural Kinds" (1997) and by Hilary Kornblith in his "Inductive Inference and Its Natural Ground" (1993). Sankey and Kornblith argue that the world has a basic natural kinds structure, and that this is important because it gives substance to the idea that nature is uniform. Specifically, it guarantees that certain properties are uniquely clustered. In Ellis (1965), I sought to vindicate scientific inductive practices by arguing for the virtues of conceptual and theoretical conservatism in inductive reasoning. I appealed to such considerations specifically to deal with Goodmanesque problems arising from the possibility of radical reconceptualizations of the world (Goodman, 1955). I argued there that there were very good pragmatic reasons for these forms of epistemic conservatism. Specifically, I argued (a) that theoretical involvement is a necessary condition for the possibility of rational non-demonstrative argument, and (b) that Goodmanesque conceptual revisions in a given field could easily destroy the theoretical involvement of the terms we use to describe the things that exist in this field. Therefore, to make such revisions unnecessarily is irrational because it must leave us powerless to argue inductively in the area of these revisions.

However, I had no wish disallow conceptual or theoretical revision altogether. I could only insist that it be thorough, and *prima facie* compatible with any other theories we might have about the kinds of things we think we are dealing with. But then I had to allow that the construction of radical alternative theoretical frameworks might not be all that difficult, especially if there were no constraints other than empirical ones on the kinds of theories that could be developed. The supposed virtues of epistemic conservatism could thus quickly lose their appeal. Theoretical involvement may be a necessary condition for the possibility of rational non-demonstrative argument. But perhaps Goodmanesque theories could easily be developed to replace the standard ones, and hence serve as a basis for Goodmanesque inferences.

From the viewpoint of scientific essentialism, the argument for theo-

retical and conceptual conservatism can be advanced a step or two further. For if the laws of nature are not imposed on the world, but arise from
the essential natures of things in the world, then it is metaphysically impossible for things to behave in any of the bizarre ways envisaged by
Goodman in his examples. For the essential nature of a thing cannot be
dependent on anything that is contingent, such as the date or place of its
existence, or whether or not, or how often, it has been observed. Consequently, the laws of nature, which derive from the dispositional properties
that things have essentially, cannot contain any references to any such contingencies. Therefore it makes a great deal of difference whether one
thinks of the laws of nature as impositions on a passive world or as arising
out of its nature. If they are imposed on a passive world, then anything
goes. But if they derive from the essential properties of things, then they
cannot be dependent on the specific circumstances of their existence.

8.10. NECESSARY CONNECTIONS BETWEEN EVENTS

Essentialism implies that some events make other events of various kinds
more or less probable. They rarely necessitate other events of quantitatively
specific kinds. Consequently, the laws of nature are mostly not deterministic, but probabilistic. However, to simplify the picture, in order to get a
clearer view of the essentialist's world, let us suppose that the causal laws
are all deterministic. That is, let us assume that events that belong to any
given causal kind always necessitate quantitatively specific kinds of events
of the corresponding effectual kind.[15] For example, let us assume that an
activating event, vis à vis any given causal power, will always necessitate
a specific kind of display of this power. Suppose, for example, that P is a natural dispositional property that would be triggered in circumstances of the
kind C to produce an effect of the kind E. Then the processes of this kind
will themselves constitute a natural kind, the essence of which is that it is
a display of P. Therefore,

L1. For all x, necessarily, if x has P, and x is in circumstances of the kind C, then
x will display an effect of the kind E, unless there are defeating conditions
that would mask this display.

Note that the necessity operator in this formula is within the scope of the
universal quantifier – that is, in *de re* position. Moreover, if a is an individual that has P necessarily – for example, because P is an essential property
of a – then we may detach and deduce that:

Necessarily, if *a* is in circumstances C, then *a* will display an effect of the kind E, unless there are defeating conditions that would mask this display.

And this, it should be noted, is a necessary connection between events of the kind that Hume rejected.

The same reasoning applies generally to all causal laws. For all such laws describe natural kinds of processes that are the displays of causal powers. The essentialist's world is therefore not one in which all events are loose and separate. On the contrary, it is a world dominated by causal powers in which events activating these powers necessitate other events that are their displays. If it were a deterministic world, as I have imagined, then any two things that are intrinsically the same must of metaphysical necessity be disposed to behave in the same way, and indeed must behave in the same way in the same circumstances. Even in an indeterministic world, something akin to this is true. Two things that are intrinsically the same must of metaphysical necessity be intrinsically disposed to behave in the same kind of way (with a real probability distribution ranging over the possible expressions of their common intrinsic properties).

The essentialist's world is therefore a bound and connected world. If what we take to be the same natural kind of thing recurs, and we do something of the same sort to it, then we should expect it to respond as any member of that kind must respond qua member of that kind. Specifically, it should display the essential dispositional properties of things of that kind for which the action we took is a trigger. If it does do so, then there is nothing to explain except perhaps how the process works. If it does not do so, then the question arises: Why should this thing be different from other things of its kind? There are many possibilities: Either (a) the thing does not belong to the natural kind to which it appears to belong (it might, for example, be a different species of the same generic kind), or (b) what we did to it was not an effective trigger (it did not belong to the appropriate natural kind of activating events), or (c) we were mistaken about what the essential properties of the kind are, or (d) the expected effect did occur, but was masked by other events, or. . . .

So, for an essentialist, the problem of induction has a rather different flavor. It is not a question of justifying the inference from "all observed As are Bs" to "all As are Bs." This inference would be justified automatically if we had good reason to believe that the As we had observed belonged to a natural kind and that the property of being a B was an essential property of the As. In that case, the problem would be rather to explain the failure of such an inference. Where we have a case of inductive failure, many possible explanations must be sorted through, and the scientific task is to do this, eliminate alternatives, and determine which of the remaining alternatives provides the best explanation. We might decide that the class designated by "A" is not a natural kind class, or that there are no essential properties of As in virtue of which they are Bs, so that if an A is a B, then this is just an accident, or that the apparent exceptions are either not really As, or they really are Bs (that is, monster-barring and monster-adjustment), or any of a number of other things. But whatever we decide, we will have learned something from our experience, and our conceptualization of the world will have been improved.

8.11. THE NEW CASE FOR EPISTEMIC CONSERVATISM

In Ellis (1965), I argued that "theoretical involvement [of the subject-matter of our inferences] is a necessary condition for the possibility of rational non-demonstrative argument" (p. 296). That being the case, I argued, it is necessary either (a) to predict that future observations will conform to accepted theories about the nature of the subject-matter, or (b) to recast our theories about the subject-matter in ways that are no less acceptable, given past observations, and to predict that future observations will conform to these new theories. No other strategy will guarantee the preservation of the kind of theoretical involvement necessary for rational non-demonstrative argument.

Without theoretical involvement, I argued, every prediction about the future or unobserved past can be justified with reference to any plausible inductive rule. For different ways of conceptualizing what is known to have happened lead to substantially different predictions about what will happen or what must have happened. Therefore, if there is no preferred way of conceptualizing what has happened, then there is no preferred way of projecting what will happen. Anything goes. Therefore, I argued, it is irrational to project any sequence of events into the future or unobserved past in such a way that if the projection were actually to be confirmed, then our theoretical understanding of the subject matter of our inference would be destroyed. This is what is wrong with Goodman's "grue" and "bleen" projections, for example. If green things like emeralds, and blue things like sapphires, really did turn out to be grue and bleen, respectively, and this were confirmed, not only by observation, but photographically and spectrographically, then our theoretical understanding of matter and of light and color would be thrown into turmoil. It would leave us in a state of bewilderment, no longer knowing what to expect about the colors of things in the future. Therefore, to accept such an inference is irrational. To do so is to reject our previous theoretical understanding of the subject-matter of the inference, not on the basis of anything that is known to have happened, but solely on the basis of what is projected will happen. If this were a rational procedure, I argued, then all theories could be rejected out of hand, and then we really would be in the a-theoretical position that Hume imagined us to be.

I see no reason to disagree with any of this. But at the time of writing the 1965 paper, I had not seen the connection with natural kinds. Nor would I have quarreled with Hume's view that anything that is conceivable (or imaginable) is possible. So I had no difficulty with the possibility of things behaving in irregular, or even bizarre, ways. For me, possibility (in this context) was just epistemic possibility. I had no conception of real possibility. Consequently, what I thought was possible would have allowed my imagination free rein. My constraints on irregular or bizarre projections of observed regularities did not derive from a belief that such projections are not real possibilities. It was just that we had no viable alternative to the kind of epistemic conservatism I then advocated. It was that or the *bush*. My residual worry was that "alternative, but so far equally satisfactory theories" about the subject-matters of our inductive inferences might turn out to be fairly easy to construct. If so, then nothing would have been gained by my argument for epistemic conservatism. For a world that is, as far as we know, compatible with many different, but overall coherent, readily constructible, and equally satisfactory, conceptions of reality, would be a world with too many choices. It would be a world with too many signposts. If alternative, Goodmanesque, global theories of reality could easily be constructed, then alternative, Goodmanesque, projections of observed regularities could be made without violating any of the requirements of epistemic conservatism.

From my present scientific essentialist perspective, my earlier worry about the possibility of alternative conceptualizations of reality now appears to be unfounded. When I was writing then, I assumed, as nearly everyone else did, that the laws of nature are contingent, and are superimposed on an unsuspecting, and passive, world. Consequently, I had no argument against the possibility of there being Goodmanesque laws of nature, and hence Goodmanesque dispositional properties. I now think that this is all wrong. The laws of nature are not contingent, but necessary, and they are not superimposed on things, but are immanent in them. The principal scientific task, according to a scientific essentialist, is to discover what natural kinds of things there are, and what their essential properties and structures are. When we know this, we will know how things are necessarily disposed to behave and interact, and thus what the laws of nature are.

The conceptualization that has resulted from this endeavor is the one in current use. Therefore, to reject it is to reject one of the principal achievements of science. For there is no alternative conceptualization that is compatible with these achievements. If there were, then we should all have heard about it. It is all very well to imagine that we might all be brains in vats, and that our scientific knowledge is consequently illusory, or that with diligence, we could redescribe the world that is so far known to us in a Goodmanesque way. But the first is a skeptical position that has no bearing on the problem of induction, and the second is an inference drawn from the alleged arbitrariness of linguistic conventions. Neither tells us anything about what the world is really like. Only our science tells us that. Hence the kind of epistemic conservatism that is required to justify our scientific inductive practices is not a stance that needs any independent justification. Epistemic conservatism of the kind required to justify our scientific inductive practices is just the determination to hold on to the achievements of science, including its conceptualizations, until we are forced to make changes to accommodate new information. If we are not epistemic conservatives in this sense, then we do not believe in the results of science, and there is nothing for us to justify.

From the standpoint of a scientific essentialist, date- or observation-dependent properties such as grue and bleen cannot be characteristic of kinds. For the date of an occurrence concerning an object of a given kind, say A, depends accidentally on its relationship to those events that serve as reference points for dating purposes. Therefore, there cannot be an essential property of As that is date-dependent – that is, there cannot be an essential property of As whose mode of operation is a function of the date. Therefore there cannot be a law of nature describing how As are by nature disposed to behave that is date-dependent. Therefore it is impossible, metaphysically impossible, that anything should be by nature grue or bleen. Such properties simply cannot exist in any world in which the laws of nature depend on the essential properties and structures of natural kinds of things.

Scientific essentialism thus imposes very strong limitations on the kinds of conceptual and theoretical innovations that can be made. It is one thing to be constrained only by considerations of empirical adequacy, and an ill-defined notion of regularity. It is another to be constrained by the requirement that things must always be supposed to behave strictly according to their natures.

If scientific essentialism is accepted, philosophy must change direction. In metaphysics, it must focus on reality rather than on language or the visual image. For all of the important concepts of essentialist metaphysics are grounded in reality, not in how we may talk about or imagine reality to be. The important modalities for metaphysics are the *de re* modalities because epistemic possibilities and necessities are grounded in ignorance rather than knowledge. If something is epistemically possible, then this is only because we have no sufficient warrant for believing that it is really impossible. If it is epistemically necessary, then it is either analytic or formally logically necessary, or known to be metaphysically necessary, but it might also be necessary because, unknown to us, it is really impossible for it to be false. If something is epistemically probable, then this is because our reasons for believing it seems stronger than those for disbelieving it. But if reality is our focus, then none of these epistemic concepts is of primary importance. What matters is how the world is, and this is for our science to tell us.

The epistemic concepts become the main concern only when we step away from reality to think about our thinking about it. They are of primary importance for the theory of rationality, for example, where what is at issue are questions about the coherence and explanatory adequacy of our belief systems. Logic, as I have argued elsewhere (Ellis, 1979), should be regarded as the theory of rational belief systems rather than the theory of truth or truth preservation. Scientific essentialism must also lead to a turning away from semantic analysis as a fundamental tool for the pursuit of metaphysical aims. For semantics is concerned with the relationships between words and the world, and so how we think about and represent the world in language. But there is no reason to think that the language we speak accurately reflects the kind of world we live in. From the fact that something is not evidently self-contradictory, for example, we are likely to conclude that it is really possible. Nevertheless, this may not be so. Our everyday language just reflects our naive judgments of what could or could not occur in the world, not judgments based on a scientific understanding of things.

Analytic and formal logical truths are necessarily true. I have no quarrel with this. But not all necessary truths are either analytic or formal logical. Most metaphysically necessary truths, for example, are synthetic since they are not true in virtue of the meanings of words. The grounds of their truth lie in the natures of the kinds of things to which reference is made. Consequently, there must be many truths that are at once synthetic, necessary, and *a posteriori*. The proposition that helium has atomic number two, for example, is not analytic since its truth does not depend on the meanings of words; it is necessary, since nothing could have atomic number two without being a species of helium, and it is *a posteriori* since it had to be established empirically that the element responsible for the mysterious Fraunhofer lines in the solar spectrum has atomic number two. Scientific essentialism thus requires that philosophers distinguish clearly between semantic issues, epistemological issues, and ontological issues.

Realistic analysis is the kind of analysis that should result when these and other fundamental distinctions are clearly recognized, and rigorously maintained – distinctions between properties, and predicates, natural kinds and arbitrary classes, species and instances, causal processes and mere sequences of events, essential, accidental and incidental properties, and so on. That is, the program of analysis should pay due attention to all of the important distinctions between kinds of objects, properties and processes existing in the world, their instantiations, and their metaphysical interconnectedness. It is not good enough to treat properties or natural kinds as sets of abstract objects or to think and reason about them as if they were nothing more than this. Realistic analyses that keep track of the universals as well as their instances are required.

I envisage, for example, the development of a realistic logic of chemistry that will not only explain instance reasoning, but do something to help us sort out the complex reasoning processes (such as those described in Section 4.6) that were involved in the early development of chemical theory. For the history of chemistry provides an excellent case study in natural kinds reasoning. The chemical elements and compounds constitute the most readily accessible system of natural kinds of substances, their properties are mostly their essential properties, and the processes they undergo in chemical interactions are all natural kinds of processes that display the essential properties of the substances involved. The question is: How should we reason about and develop our knowledge of such a system? A logic of chemical theory that did justice to the complexities of the subject-matter and that chemists themselves would find useful would be a significant step in the right direction.

NOTES

1. I am of course aware that there are other theories about laws of nature than the two that are contrasted here. There is, for example, the conventionalist theory, according to which the laws of nature are conventions adopted for their utility in or-

292

ganizing or structuring our knowledge. On this theory, laws of nature are, strictly speaking, neither true nor false, and if, as a *façon de parlez*, we should say that they are true, then we can mean only that they are the best for this purpose.

2. For example, by J.J.C. Smart (1993) in his paper, "Laws of Nature as a Species of Regularities."

3. Bas van Fraassen (1989) calls this "the identification problem," because for him the problem is to identify the relationship in virtue of which a universal generalization acquires nomic status. However, this is like looking for a relationship that will convert a material conditional into a subjunctive conditional, and this is just the wrong way of looking at it. Subjunctive conditionals are essentially modal; they are not material conditionals that have somehow acquired modality.

4. The point of "Causal Powers and Laws of Nature" (Ellis, 1999a) was to argue that the scientific essentialist can offer a better perspective on causal laws than a Humean theorist can, and to defend the essentialist theory against likely objections. Specifically, my aim was to show that an essentialist can explain (a) what the causal laws of nature are (descriptions of natural kinds of processes), (b) what their truth-makers are (the causal powers, capacities, and propensities of the objects necessarily involved in these processes), and (c) what makes these laws necessary (because things of the kinds involved must, in appropriate circumstances, display the causal powers, and so on on which their identities depend), and to argue that (d) these theses can all be defended adequately against the most common objections.

5. The criticisms I have to make of Humeanism were developed in other ways in Caroline Lierse's (1996) paper.

6. First-order logic, for example, is a logical theory arising out of a Humean worldview. It is not a logical theory that can readily be adapted to reasoning about universals, or about necessary connections in nature. I do not say that more elaborate logical systems that could do the job adequately cannot be constructed. It is just that a lot of work would need to be done in logical theory to accommodate either of these non-Humean theories of laws of nature.

7. This claim was made by d'Alembert in 1743 in his *Traité de Dynamique*. For an English translation of the relevant passage, see Magie (1935, 55–8). Laudan (1968) argues that by taking an even-handed stance on the issue of the true measure of force, and arguing that the dispute was really only about words, D'Alembert (1743) did not finally settle the *vis viva* controversy, as many have supposed. For very few natural philosophers of the period thought that the dispute was just about words. It is true that the intensity of the controversy diminished in the decades following D'Alembert's publication. Nevertheless, the controversy did continue throughout the century, and if there was a developing consensus that the dispute was dead, this was only because the supporters of *vis mortua* (momentum) as the true measure of force believed that they had won the argument.

8. The case for this is powerfully argued in Iltis (1973). Iltis argues that the Leibnizian-Newtonian controversy reflected in the Clarke-Leibniz correspondence "was fundamentally a clash of philosophical world views on the nature of God, matter, and force" (p. 343). "The Newtonian and Leibnizian groups of the 1720s developed a commitment to the mother scheme and took on the task of defending that system against the perceived threats of outside attacks. ...They were unwilling and *unable* to see that the other side had valid arguments. The early *vis viva* controversy

of the 1720s was therefore the result of a problem of communication brought about by the inability of the participants to cross the boundary lines of their particular natural philosophies" (p. 344).

9. See Laudan (1968) for examples.

10. In fact, Sir Henry Cavendish did explore the possibility of a dynamical theory of heat, based on Leibniz's concept of *vis viva*, in the late eighteenth century, and wrote a draft treatise on the subject. Unfortunately, Cavendish did not finish rewriting his draft, and the treatise was not published in his lifetime. It was buried amongst other papers, and did not surface again until the late 1960s. See McCormmach (1988).

11. Indeed this view is still widely held. "Post-Humean empiricists have shared Edna St. Vincent Millay's view of the world as one damn thing after another" (Levin, 1987, 288).

12. Roger Boscovich was another eighteenth-century natural philosopher to reject the thesis of the essential passivity of matter. Boscovich is especially interesting in this context because he made causal powers central to his ontological theory and rejected the theory that matter consists ultimately of impenetrable, extended atoms of various shapes and sizes, arranged or moving in various ways. Boscovich conceived of matter as consisting entirely of centers of attractive and repulsive forces. His atoms were just mobile points in space that had the power of attracting or repelling other such points according to a certain universal law. But Boscovich was a long way ahead of his time. Even today, belief in the fundamental passivity of matter predominates in Anglo-American philosophy.

13. See Ellis (1990, Part II).

14. This does not worry me, for reasons to be explained shortly. Nor should it worry Jackson, who bit a similar bullet some time ago when he argued that conditional probability does not measure the probability of truth of an indicative conditional, but the degree to which it is assertible (Jackson, 1979). The probability of its truth, he said, is the probability of the corresponding material conditional – a claim that can be shown to have some strongly counter-intuitive consequences (Ellis, 1984). But see also Jackson (1984).

15. See Section 3.9 for an explanation of these terms.

Bibliography

Achinstein, P., 1974, "The Identity of Properties," *American Philosophical Quarterly* 11, 257–76.

Ackerman, D.F., 1986, "Essential Properties and Philosophical Analysis," *Midwest Studies in Philosophy* 11, 315–30.

Anderson, E., 1996, "Generalizing Scientific Essentialism," *Victorian Centre for the History and Philosophy of Science Preprint Series* 96/1, La Trobe University, Department of Philosophy.

Armstrong, D.M., 1968, *A Materialist Theory of the Mind*. London, Routledge and Kegan Paul.

——, 1973, *Belief, Truth and Knowledge*. Cambridge, Cambridge University Press.

——, 1978, *Universals and Scientific Realism*, 2 vols. Cambridge, Cambridge University Press.

——, 1983, *What Is a Law of Nature?* Cambridge, Cambridge University Press.

——, 1989, *Universals: An Opinionated Introduction*. Boulder, San Francisco, and London, Westview Press.

——, 1997, *A World of States of Affairs*. Cambridge, Cambridge University Press.

——, 1999a, "Reply to Ellis," in H.Sankey, ed., *Causation and Laws of Nature*, pp. 43–48.

——, 1999b, "The Causal Theory of Properties: Properties According to Shoemaker, Ellis and Others," *Philosophical Topics*, 26, 25–37.

Averill, E.W., 1990, "Are Physical Properties Dispositions?" *Philosophy of Science* 57, 118–32.

Ayers, M.R., 1981, "Locke versus Aristotle on Natural Kinds," *The Journal of Philosophy* 78, 247–72.

Backhouse, R.E., 1992, "Lakatos and Economics." In *Perspectives in the History of Economic Thought*, ed. by S.T. Lowry, 19–34. Aldershot and Brookfield, VT, Edward Elgar.

——, 1993, "Lakatosian Perspectives on General Equilibrium Analysis," *Economics and Philosophy* 9, 271–82.

Bealer, G., 1987, "The Philosophical Limits of Scientific Essentialism," in J.Tomberlin, ed., *Philosophical Perspectives* 1, Atascadero, CA, Ridgeway, 289–365.

——, 1994, "Mental Properties," *The Journal of Philosophy* 91, 185–208.

Bhaskar, R., 1978, *A Realist Theory of Science*. Hassocks, Sussex, Harvester Press.

Bigelow, J.C., 1990, "The World Essence," *Dialogue*, 29, 205–17.

——, 1999, "Scientific Ellisianism," in H. Sankey, ed., *Causation and Laws of Nature*, 56–76.

Bigelow, J.C. and Pargetter, R.J., 1988, "Quantities," *Philosophical Studies* 54, 75–92.

——, 1989, "A Theory of Structural Universals," *Australasian Journal of Philosophy* 67, 1–11.

——, 1990, *Science and Necessity*. Cambridge, Cambridge University Press.

Bigelow, J.C., Ellis, B.D. and Lierse, C., 1992, "The World as One of a Kind: Natural Necessity and Laws of Nature," *British Journal for the Philosophy of Science* 43, 371–88.

Boyd, R., 1991, " Realism, Anti-foundationalism and the Enthusiasm for Natural Kinds," *Philosophical Studies* 61, 127–48.

Brody, B.A., 1967, "Natural Kinds and Real Essences," *The Journal of Philosophy* 64, 431–46.

Burtt, E.A., 1932, *Metaphysical Foundations of Modern Science,* 2nd ed. London, Routledge and Kegan Paul.

Butts, R.E., 1977, "Consilience of Inductions and the Problem of Conceptual Change in Science," in R.G. Colodny, ed., *Logic, Laws and Life*. Pittsburgh, Pittsburgh University Press, 71–88.

Campbell, K.K., 1990, *Abstract Particulars*. Oxford, Blackwell.

Caplan, A., 1981, "Back to Class: A Note on the Ontology of Species," *Philosophy of Science* 48, 130–40.

Carroll, J.W., 1994, *Laws of Nature*. Cambridge, Cambridge University Press.

Cartwright, N., 1983, *How the Laws of Physics Lie*. Oxford, Oxford University Press.

Chalmers, A.F., 1987, "Bhaskar, Cartwright and Realism in Physics," *Methodology in Science* 20, 77–96.

Code, A., 1986, "Aristotle, Essence and Accident," in R.E. Grandy, R. Warner, eds., *Philosophical Grounds of Rationality,* Oxford, Oxford University Press, 411–39.

Costa, M.J., 1989, "Hume and Causal Realism," *Australasian Journal of Philosophy* 67, 172–190.

D'Alembert, J. le R., 1743, Extract from *Traité de Dynamique,* trans. W.F. Magie in W.F. Magie, ed. *A Source Book in Physics*. New York and London, McGraw-Hill, 1935, 55–8.

Davies, M.K. and Humberstone, I.L., 1980, "Two Notions of Necessity," *Philosophical Studies* 38, 1–30.

De Sousa, R., 1984, "The Natural Shiftiness of Natural Kinds," *Canadian Journal of Philosophy* 14, 561–80.

Donnellan, K.S., 1983, "Kripke and Putnam on Natural Kind Terms," in C. Ginet, S. Shoemaker, eds., *Knowledge and Mind*. New York and Oxford, Oxford University Press, 84–104.

Dowe, P., 1999, "Good Connections: Causation and Causal Processes," in H. Sankey, ed., *Causation and Laws of Nature,* 247–64.

Dretske, Fred I., 1977, "Laws of Nature," *Philosophy of Science* 44, 248–68.

Duhem, P., 1954, *The Aim and Structure of Physical Theory,* trans. P.P. Wiener from French 1914 edition. Princeton, Princeton University Press.

Dupre, J., 1993, *The Disorder of Things: Metaphysical Foundations of the Disunity of Science.* Cambridge, MA, and London, Harvard University Press.

Elder, C.L., 1989, "Realism, Naturalism, and Culturally Generated Kinds," *The Philosophical Quarterly* 39, 425–44.

——, 1992, "An Epistemological Defense of Realism about Necessity," *The Philosophical Quarterly* 42, 317–36.

——, 1994a, "Higher and Lower Essential Natures," *American Philosophical Quarterly* 31, 255–265.

——, 1994b, "Laws, Natures and Contingent Necessities," *Philosophy and Phenomenological Research* 54, 649–67.

——, 1995, "A Different Kind of Natural Kind," *Australasian Journal of Philosophy* 73, 516–31.

Ellis, B.D., 1965, "A Vindication of Scientific Inductive Practices," *American Philosophical Quarterly* 2, 296–305.

——, 1976, "The Existence of Forces," *Studies in the History and Philosophy of Science* 7, 171–85.

——, 1979, *Rational Belief Systems.* Oxford, Blackwell.

——, 1984, "Two Theories of Indicative Conditionals," *Australasian Journal of Philosophy* 62, 50–66.

——, 1985, "What Science Aims to Do," in P.M. Churchland, C.A. Hooker, eds., *Images of Science.* Chicago, Chicago University Press, 48–74.

——, 1987, "The Ontology of Scientific Realism," in P. Pettit, R. Sylvan, J. Norman, eds., *Metaphysics and Morality: Essays in Honour of J.J.C. Smart.* Oxford, Blackwell, 50–70.

——, 1990, *Truth and Objectivity.* Oxford, Blackwell.

——, 1992a, "Idealization in Science," in C. Dilworth, ed., *Idealization IV: Intelligibility in Science, Poznan Studies in the Philosophy of the Sciences and the Humanities* 26. Amsterdam and Atlanta GA, Rodopi.

——, 1992b, "Scientific Platonism," essay review of *Science and Necessity,* J.C. Bigelow and R.J. Pargetter, 1991, *Studies in the History and Philosophy of Science* 23, 665–79.

——, 1996, "Natural Kinds and Natural Kind Reasoning," in P. Riggs, ed., *Natural Kinds, Laws of Nature and Scientific Methodology,* 11–28. Dordrecht, Kluwer.

——, 1999a, "Causal Powers and Laws of Nature," in H. Sankey, ed. *Causation and Laws of Nature,* 21–42.

——, 1999b, "Bigelow's Worries about Scientific Essentialism," in H. Sankey, ed., *Causation and Laws of Nature,* 77–97.

——, 1999c, "Response to David Armstrong," in H. Sankey, ed., *Causation and Laws of Nature,* 49–55.

——, 1999d, "A World of States of Affairs," by D.M. Armstrong, *Metascience* 8, 63–73.

——, forthcoming, "The New Essentialism and the Scientific Image of Mankind," *Epistemologia.*

Ellis, B.D. and Lierse, C.E., 1994a, "Dispositional Essentialism," *Australasian Journal of Philosophy* 72, 27–45.

——, 1994b, "The Fundamental Importance of Natural Kinds." La Trobe University, *Victorian Centre for the Philosophy of Science Preprint Series* 3/94.

Euler, L., 1795, *Letters of Euler to a German Princess,* 2 vols. Trans. from French by Henry

Hunter. Reprinted with a new Introduction by Andrew Pyle, Bristol, England, Thoemmes Press, 1997.

Fales, E., 1979, "Relative Essentialism," *British Journal for the Philosophy of Science* 30, 349–70.

——, 1982, "Natural Kinds and Freaks of Nature," *Philosophy of Science* 49, 67–90.

——, 1986, "Essentialism and the Elementary Constituents of Matter," *Midwest Studies in Philosophy* 11, 391–402.

Fales, E., 1990, *Causation and Universals*. London, Routledge and Kegan Paul.

Farrington, B., 1951, *Francis Bacon: Philosopher of Industrial Science*. London, Lawrence and Wishart.

Fine, K., 1987, "Essence and Modality," in J. Tomberlin, ed., *Philosophical Perspectives* 8, 1–16.

——, 1994/5, "Ontological Dependence," *Proceedings of the Aristotelean Society* XCV, 269–89.

——, 1995, "The Logic of Essence," *Journal of Philosophical Logic* 24, 241–73.

Forster, M.R., 1988, "Unification, Explanation, and the Composition of Causes in Newtonian Mechanics," *Studies in the History and Philosophy of Science* 19, 55–101.

Fox, J.F., 1987, "Truthmaker," *Australasian Journal of Philosophy* 65, 188–207.

Freddoso, A.J., 1986, "The Necessity of Nature," *Midwest Studies in Philosophy* 11, 215–42.

Friedman, M., 1953, *Essays in Positive Economics*. Chicago, University of Chicago Press.

Fukuyama, F., 1995, *Trust: The Social Virtues and the Creation of Prosperity*. London, Hamish Hamilton.

Geach, P., 1969, *God and the Soul*. London, Routledge and Kegan Paul.

Ghiselin, M., 1974, "A Radical Solution to the Species Problem," *Systematic Zoology* 23, 536–44.

——, 1987, "Species Concepts, Individuality, and Objectivity," *Biology and Philosophy* 2, 127–44.

Goodman, N., 1955, *Fact, Fiction and Forecast*. Cambridge, MA, Harvard University Press.

Hacking, I., 1991, "A Tradition of Natural Kinds," *Philosophical Studies* 61, 109–26.

Hall, A.R., 1954, *The Scientific Revolution 1500–1800*. London, New York, and Toronto; Longmans, Green.

Hands, D.W., 1984, "What Economics Is Not: An Economist's Response to Rosenberg," *Philosophy of Science* 51, 495–503.

Hardegree, G.M., 1982, "An Approach to the Logic of Natural Kinds," *Pacific Philosophical Quarterly* 63, 122–32

Harper, W., 1989, "Consilience and Natural Kind Reasoning," in J.R. Brown, J. Mittelstrass, eds., *An Intimate Relation*. Dordrecht, Kluwer, 115–52.

Harré, R., 1970, "Powers," *British Journal for the Philosophy of Science* 21, 81–101.

Harré, R. and Madden, E.H., 1973, "Natural Powers and Powerful Natures," *Philosophy* 48, 209–30.

——, 1975, *Causal Powers: A Theory of Natural Necessity*. Oxford, Blackwell.

Hull, D.L., 1976, "Are Species Really Individuals?" *Systematic Zoology* 25, 174–91.

——, 1978, "A Matter of Individuality," *Philosophy of Science* 45, 335–60.

——, 1981, "Kitts and Kitts and Caplan on Species," *Philosophy of Science* 48, 141–52.

Hume, D., 1777, *Enquiries Concerning Human Understanding and Concerning the Princi-*

298

ples of Morals, 3rd ed. Eds L.A. Selby-Bigge, P.H. Nidditch. Oxford, Clarendon Press, 1975.

Hutton, W., 1996, *The State We're In.* London, Vintage Books.

Iltis, C., 1973, "The Leibnizian-Newtonian Debates: Natural Philosophy and Social Psychology," *The British Journal for the History of Science* 6, 343–77.

Jackson, F.C., 1979, "On Assertion and Indicative Conditionals," *Philosophical Review* 88, 565–79.

——, 1984, "Two Theories of Indicative Conditionals: A Reply to Brian Ellis," *Australasian Journal of Philosophy* 62, 67–76.

Kant, I., 1787, *Critique of Pure Reason,* 2nd ed., trans. Norman Kemp Smith. London, Macmillan, 1950.

Kim, J., 1973, "Causation, Nomic Subsumption and the Concept of Event," *The Journal of Philosophy* 70, 217–36.

——, 1982, "Psychological Supervenience," *Philosophical Studies* 41, 51–70.

Kornblith, H., 1993, *Inductive Inference and its Natural Ground.* Cambridge, MA, MIT Press.

Kripke, S., 1972, "Naming and Necessity," in D. Davidson, G. Harman, eds., *Semantics of Natural Language.* Dordrecht, Reidel, 252–355.

Lakatos, I., 1970, "Falsification and the Methodology of Scientific Research Programmes," in I. Lakatos, and A. Musgrave, eds., *Criticism and the Growth of Knowledge.* Cambridge, Cambridge University Press, 91–196.

Langton, R. and Lewis, D.K., 1998, "Defining "Intrinsic"," *Philosophy and Phenomenological Research* 58, 333–46.

Laudan, L.L., 1968, "The *Vis Viva* Controversy: A Postmortem," *Isis* 59, 131–43.

Lavoisier, A., 1789, *Traité Elémentaire de Chemie.* Book 1, trans. R. Kerr, published as *Elements of Chemistry,* Chicago, Illinois, Henry Regnery, for the Great Books Foundation, 1949.

Levin, M., 1987, "Rigid Designators: Two Applications," *Philosophy of Science* 54, 283–94.

Lewis, D.K., 1973, *Counterfactuals.* Cambridge, MA, Harvard University Press.

——, 1983a, "Extrinsic Properties," *Philosophical Studies* 44, pp. 197–200.

——, 1983b, "New Work for a Theory of Universals," *Australasian Journal of Philosophy* 61, 343–77.

——, 1986a, "Against Structural Universals," *Australasian Journal of Philosophy* 64, 25–46.

——, 1986b, *On the Plurality of Worlds.* Oxford, Blackwell.

——, 1994, "Humean Supervenience Debugged," *Mind* 412, 473–90.

Lierse, C.E., 1996, "The Jerrybuilt House of Humeanism," in P.J. Riggs, ed., *Natural Kinds, Laws of Nature and Scientific Methodology,* 29–48. Dordrecht, Kluwer.

Locke, J., 1690, *An Essay Concerning Human Understanding,* P.H. Nidditch, ed. New York and Oxford, Oxford University Press, 1975.

Mackie, J.L., 1977, "Dispositions, Grounds and Causes," in R. Tuomela, ed., *Dispositions.* Dordrecht, Reidel, 99–107.

Macnamara, J, 1991, "Understanding Induction," *British Journal for the Philosophy of Science* 42, pp 21–48.

Martin, C.B., 1994, "Dispositions and Conditionals," *The Philosophical Quarterly* 44, 1–8.

Mayer, J.R., 1842, "Remarks on the Forces of Inorganic Nature," trans. G.C. Foster in
W.F. Magie, ed., *A Source Book in Physics*. New York and London, McGraw-Hill,
1935, 196–203.

McCormmach, R., 1988, "Henry Cavendish on the Theory of Heat," *Isis* 79, 37–67.

McCullagh, C.B., 1998, *The Truth of History*. London, Routledge and Kegan Paul.

Mellor, D.H., 1974, "In Defense of Dispositions," *The Philosophical Review* 83, 157–81.

——, 1982, "Counting Corners Correctly," *Analysis* 42, 96–7.

Molnar, G., 1999, "Do Not Multiply Necessities Beyond Necessity." Unpublished
manuscript.

Mumford, S.D., 1994, "Dispositions, Supervenience and Reduction," *The Philosophical
Quarterly* 44, 419–38.

——, 1995a, "Dispositions, Bases, Overdetermination and Identities," *Ratio* 8, 42–62.

——, 1995b, "Ellis and Lierse on Dispositional Essentialism," *Australasian Journal of Philosophy* 73, 606–12.

Nelson, A., 1990, "Are Economic Kinds Natural?" *Minnesota Studies in the Philosophy
of Science* XIV, 101–35.

Nelson, A., 1992, "Human Molecules," in N. de Marchi, ed. op. cit., 113–33, with com-
mentary by B. Caldwell, 135–49, and reply by A. Nelson, 151–54.

Pargetter, R.J. and Prior, E.W., 1982, "The Dispositional and the Categorical," *Pacific
Philosophical Quarterly* 63, 366–70.

Poincaré, H., 1905, *Science and Hypothesis*. Re-publication of first English edition, New
York, Dover. 1952.

Popper, K.R., 1959, *The Logic of Scientific Discovery*. London, Hutchinson.

——, 1962, "The Propensity Interpretation of the Calculus of Probability, and the
Quantum Theory," in S. Korner, ed., *Observation and Interpretation in the Philosophy of Physics*. New York, Dover.

Price, H.H., 1969, *Thinking and Experience*. Second Edition; London, Hutchinson University Library.

Price, Huw, 1999, "The Role of History in Microphysics," in H. Sankey, ed., *Causation and Laws of Nature*, 331–46.

Prior, E.W., 1982, "The Dispositional/Categorical Distinction," *Analysis* 42, 93–6.

——, 1985, *Dispositions*. Aberdeen, Aberdeen University Press.

Prior, E.W., Pargetter, R.J. and Jackson, F.C., 1982, "Three Theses about Dispositions,"
American Philosophical Quarterly 19, 251–7.

Quinton, A., 1957, "Properties and Classes," *Proceedings of the Aristotelian Society* LVIII,
33–58.

Rappaport, S., 1995, "Is Economics Empirical Knowledge?" *Economics and Philosophy*
11, 137–58.

Reeder, N., 1995, "Are Physical Properties Dispositions?," *Philosophy of Science* 62,
141–9.

Reichenbach, H., 1958, *The Philosophy of Space and Time*. New York; Dover. English
trans. M. Reichenbach, J. Freund of Reichenbach's *Philosophie der Raum-Zeit-
Lehre*.

Rosenberg, A., 1986, "Lakatosian Consolations for Economics," *Economics and Philosophy* 2, 127–39.

——, 1992, *Economics – Mathematical Politics or Science of Diminishing Returns*. Chicago,
University of Chicago Press.

Ruby, J., 1986, "The Origins of Scientific "Law," *Journal of the History of Ideas* 47, 341–59.

Ruse, M., 1987, "Biological Species: Natural Kinds, Individuals, Or What?" *British Journal for the Philosophy of Science* 38, 225–42.

Salmon, N., 1981, *Reference and Essence*. Princeton, Princeton University Press.

Salmon, W.C., Jeffrey, R.C., and Greeno, J.G., 1971, *Statistical Explanation and Statistical Relevance*. Pittsburgh, University of Pittsburgh Press.

Salmon, W.C., 1984, *Scientific Explanation and the Causal Structure of the World*. Princeton, Princeton University Press.

Sankey, H., 1997, "Induction and Natural Kinds," *Principia* 1, 235–54.

——, 1999, *Causation and Laws of Nature*. Dordrecht, Kluwer.

Sawyer, K., Beed, C., Sankey, H., and Ellis, B., 1998, "Metaeconomic Principles." Paper read at Western Division meeting of American Philosophical Association.

Schlesinger, G., 1963, *Method in the Physical Sciences*. London, Routledge and Kegan Paul.

Schwartz, S.P., ed., 1977, *Naming, Necessity and Natural Kinds*. Ithaca, NY, Cornell University Press.

——, 1980, "Natural Kinds and Nominal Kinds," *Mind* 89, 186–8.

Scriven, M., 1961, "The Key Property of Physical Laws – Inaccuracy," in H. Feigl, G. Maxwell, eds., *Current Issues in the Philosophy of Science*. New York, Holt, Rinehart and Winston, 91–104.

Shoemaker, S., 1980, "Causality and Properties," in P. van Inwagen, ed., *Time and Cause: Essays Presented to Richard Taylor*. Dordrecht, Reidel, 109–35.

Smart, J.J.C., 1963, *Philosophy and Scientific Realism*. London, Routledge and Kegan Paul.

——, 1987, "Replies," in P. Pettit, R. Sylvan, J. Norman, eds., *Metaphysics and Morality: Essays in Honour of J.J.C. Smart*. Oxford, Blackwell, 173–95.

——, 1993, "Laws of Nature as a Species of Regularities," in J. Bacon, K.K. Campbell, L. Reinhardt, eds., *Ontology, Causality and Mind: Essays in Honour of D.M. Armstrong*. Cambridge, Cambridge University Press, 152–69. Reply by D.M. Armstrong, *op. cit.*, 169–74.

Suchting, W., 1988, "Comments on "Bhaskar, Cartwright and Realism in Physics" by Alan Chalmers," *Methodology in Science* 21, 77–80.

Swinburne, R., 1983, "Reply to Shoemaker," in L.J. Cohen, Mary Hesse, eds., *Aspects of Inductive Logic*. Oxford, Oxford University Press, 313–20.

Swoyer, C., 1982, "The Nature of Natural Laws," *Australasian Journal of Philosophy* 60, 203–23.

Thayer, H.S., 1953, *Newton's Philosophy of Nature*. New York, Hafner.

Thomason, R., 1969, "Species, Determinates and Natural Kinds," *Noûs* 3, 95–101.

Thompson, I.J., 1988, "Real Dispositions in the Physical World," *British Journal for the Philosophy of Science* 39, 67–80.

Tobin, J., 1977, "How Dead Is Keynes?" *Economic Inquiry* 15, 459–68.

Toohey, B., 1994, *Tumbling Dice: The Story of Modern Economic Policy*. William Heinemann Australia.

Tooley, M., 1977, "The Nature of Laws," *Canadian Journal of Philosophy* 7, 667–98.

——, 1987, *Causation: A Realist Approach*. Oxford; Clarendon Press.

van Fraassen, B.C., 1980, *The Scientific Image*. Oxford, Clarendon Press.

301

——, 1989, *Laws and Symmetry.* Oxford, Clarendon Press.

Walras, L., 1874, *Elements of Pure Economics.* Trans. W. Jaffe, Homewood, Ill.; Irwin, 1954.

Weintraub, E.R., 1988, "The Neo-Walrasian Research Programme Is Empirically Progressive." In *The Popperian Legacy in Economics,* edited by N. de Marchi, ed., 213–27. Cambridge, Cambridge University Press.

Wilkerson, T.E., 1986, "Natural Kinds and Identity, A Horticultural Enquiry," *Philosophical Studies* 49, 63–69.

——, 1988, "Natural Kinds," *Philosophy* 63, 29–42.

——, 1993, "Species, Essences and the Names of Natural Kinds," *Philosophical Quarterly* 43, 1–19.

——, 1995, *Natural Kinds.* Avebury, Ashgate Publishing Co.

Windschuttle, K., 1996, *The Killing of History,* 2nd ed. Paddington, New South Wales, Mcleay Press.

Winnie, J., 1970, "Special Relativity Without One-Way Velocity Assumptions," *Philosophy of Science* 37, 81–99, 223–38.

Index

causal laws, 5, 6, 11, 52–3, 137–8,
206; and dispositions, 113; as
non-deterministic, 286–7;
inapplicability of in social
sciences, 177–8
causal powers, 4–6, 49–55: active and
passive, 107–9; as potentialities,
133–5; as relations to generic
kinds of events, 133–4, 206, 208;
as truthmakers for causal laws,
5, 154, 157, 210, 217, 222, 244,
270; identity of, 53, 133–4;
meta-, 21; quantitative nature
of, 132–3
causal processes, 49–55, 133–4, 158,
218: elementary, 51, 218, 224,
250
causal properties: 2, 9–10, 51, 52, 243;
in scientific essentialism, 4, 5–6,
48, 106, 147, 160; source of,
263–7
classes: natural kind, 101; property, 99
classification: 17–18
conditionals: acceptability conditions
for, 281–3; counterfactual, 12,
281–2; essentialist theory of, 12,
278–9; realistic theory of,
279–83; subjunctive, 113; truth
conditions for, 271, 272
co-extensionality, 86–8
conjunctive properties: defined, 90–1;
and principle of non-
proliferation, 96
conjunctive universals, 75–6
connectivity: principle of, 122, 126,
140n14
conservation laws: as convention, 214;
and idealization problem, 212
Contingency Thesis, 113, 117, 122,
123, 171
Continuing Existence Argument: for
categorical realism, 114, 116–17,
139n12

conventionalist theory of laws. See
laws of nature: conventionalist
theory of.
correspondence theory of truth,
271–2
counterfactual conditionals. See
conditionals: counterfactual.
Critique of Pure Reason (Kant), 38–9

Darwin, Charles, 2
De Sousa, R.: and definition of
natural kinds, 55–6n1
definitions: real distinguished from
nominal, 35
Descartes, René, 2, 107, 264, 265
determinables, 19, 65–7, 70, 149
determinates, 19, 65–7, 119
Difference Argument: for categorical
realism, 114, 116–17
dispositional foundationalism, 111–2,
121–2
dispositions: in Bigelow and
Pargetter's ontology, 150, 155;
and categorical realism, 112–15,
117, 118; causal basis of, 119,
124–5, 129–30; causally
indeterminant, 131; defined, 9,
112; Mellor's theory of, 111–2,
121–3; mere Cambridge, 125–6;
real, 125–6; stochastic, 129,
130–1
Distinctness Thesis, 120–1
Donnellan, K.S., 101n4
Dretske, Fred I., 210, 215
Duhem, Pierre, 158, 159
Dupré, John, 169
dynamic universals, 18, 25–6, 74, 99

economics: contrasted with physical
sciences, 178–9, 184–98;
dogmatism in, 185, 189, 195; and
empiricism, 186, 197–8;
Lakatosian methodology in, 185,

194–5; laws of, 10, 178–9; models in, 186–7; neoclassical theory of, 184

Elder, Crawford, 101n1

energy transmission processes, 249–50

essences: kind and individual, 11, 237–41; Lockean real, 31; of natural kinds, 21–2, 54–5, 98, 237–41; nominal, 32, 78, 185–6

essentiality requirement: of natural kinds, 21

Euler, Leonhard: on causal powers, 108, 109

events and processes: all, 10, 61, 204–6, 248–9, 255–6; elementary, 249; in hierarchy of natural kinds, 71–2; as members of natural kinds, 22; necessary connections between, 286–8

existential entailment, 64, 65

explanations: in biology, 167–70; causal process, 157–60; in chemistry, 161–5; covering law, 158–9; in economics, 184–90; essentialist, 10, 160–5, 165–7, 223; in history, 179–82; in physics, 165–7; positivist theories of, 159; in psychology, 170–3; realist theories of, 160, 182–4; in social sciences, 177–9

explanatory frameworks, 189–90

extensionality: requirement of in logical analysis, 270, 272

Fales, Evan, 251–2, 102n5, 138n4, 156

Fine, Kit, 102n9

forces: as essentially external, 264; Leibnizian conception of, 266; and scientific realism, 147

Forster, Malcolm, 125

Freddoso, A.J., 58n20, 257n4

Friedman, Milton, 188, 198

Geach, Peter, 126

generic kinds, 3, 32, 77

global laws, 10, 12, 204, 205

Goodman, Nelson, 285, 288, 289, 290

Hacking, Ian, 55–6n1, 104n23

Hall, A.R., 203

Hardegree, G. M., 104n24

Harré, Rom, xiii, 217, 262

Hempel's covering law model, 158, 159

hierarchies: of natural kinds, 20, 8, 61, 67–8

Hume worlds, 46, 57, 156–7, 244

Humean Supervenience thesis, 4–5, 45–6, 154, 155, 242–3, 245, 248, 261–3, 293

Humeanism, 122–3, 155–6, 209, 210; common theses of, 7, 117; compared to essentialism, 4–5, 7, 47–9, 245–8, 261–3; and idealization, 211–13; and imagined possibility, 231–234; and induction, 283; and natural necessity, 211, 229–31; ontology, 44–7

idealism: contrasted with scientific realism, 147

idealization: in science, 11, 150–4, 167, 183, 224–5; in economics, 186, 188, 189–90; in essentialist explanations, 174, 221; and model theory, 153; and processes, 218

identities: contingent, 54

image: manifest, 232

induction, 283–6, 287–8: essentialist perspective on, 289–90; and theoretical involvement requirement, 288–9

inertia: in mechanism, 108–9

infimic species: distinguished from instances, 66; in hierarchy of natural kinds, 3, 70; and quantitative universals, 8–9

inpenetrability: in mechanism, 108–9

instance: distinguished from species, 80

instantiation in: distinguished from instantiation by, 25, 73–4

intrinsicality: causal concept of, 26–9; defined, 8; logical concept of, 26; as a property of the bearing relation, 27; in theoretical models, 30

Jackson, Frank C., 119

Kant, Immanuel: on synthetic *a priori* knowledge, 38–9

Keynesian economics, 186, 197

Kim, Jaegwon, 27, 28

kinds: biological, 167–70; chemical, 3, 161–5; cluster, 32; defining properties, 76–7; global, 248–53; physical, 2, 165–7; spectral, 98, 134, 174; uninstantiated, 81–2. *See also* natural kinds.

Kirchhoff, R., 199n10

knowledge: *a priori* and *a posteriori*, 38–42

Kornblith, Hilary, 169, 285

Kripke, Saul, 54, 55, 172, 238, 257n2

Lakatosian methodology: and economics, 185–6, 194–5

Lavoisier, Antoine, 125, 164–5

laws of nature, 58n20, 163, 174; in categorical realism, 113–14; causal, 130, 206–7; classification of, 204–6; conservation 29–30; conventionalist theory of, 37, 41–2, 43–4, 137, 213–15; contrapositive formulations of,

204–5; and dispositional properties, 128; empiricist theory of, 47– 8; in essentialism, 4, 217–22; and formal principles, 225–6; global, 248–9, 255, 275; Humean theory of, 40–1, 45–7, 211–13; and idealization, 208–9, 210, 211, 213, 214, 22–5; immanence of, 289; Kantian theory of, 38–9, 41; necessity of, 6, 208, 210, 215–6, 219–21, 273–4; objectivity of, 209; ontological problem of, 209, 210, 213, 222; regularity theory of, 139n9, 211–3; scope of, 10, 204; statistical, 130–2, 206; structural,10–11; truthmakers for, 112, 128–9, 217

Leibniz, Gottfried, 108–9, 263–5

Leslie, John, 251

Lewis, David K., 151, 154,, 277

Lierse, Caroline, xiii, xiv, 44, 76, 118

Locke, John, 2, 55, 107–8, 109, 168–9

logic: as the theory of rational belief systems, 281–2; as the theory of truth preservation, 283

logical analysis, 268–72

Mach, Ernst, 159, 213

Mach's Principle, 28

Mackie, John L.: and rejection of intrinsic causal powers, 57n17

Madden, Edwin H., xiii, 217, 262

mass: whether an intrinsic or extrinsic property, 28; 58n19

maxim of discovery, 90

Mayer, J.R., 265–6

mechanism: defined, 1–2, 107

Mellor, Hugh 111, 126

micro-reduction: theory of, 64

Mill, J.S., 104n23

models: economic contrasted with scientific, 185, 190–94

multiple manifestation: principle of, 122, 126

natural kinds: in biology, 167–70; causal laws limited to, 6; of causal and stochastic processes, 125; defined, 19–23, 26, 99; dynamic, 23, 25, 61, 75; essences, 11–12; of events and processes, 3, 51; fundamental, 275–67; grid of, 145; hierarchies of, 2–3, 69–70, 145; instantiation of, 23, 73, 80; no negative, 101; of objects or substances, 3; substantive, 23, 25, 61, 75; simple, 31–2; speciation of, 20
natural necessity, 11, 12, 61; and global laws, 248–9; in Humeanism, 229–31; possible worlds theory of, 241–5, 274–8; principles of, 256–7
necessary co-extensionality: and property identity, 86–8
necessity: *a posteriori,* 219–220, 234; analytical, 180; analytical distinguished from metaphysical, 235–7; distinguished from *de dicto,* 32–3; external, 264; formal logical distinguished from analytical, 234; formal logical distinguished from natural, 11; metaphysical, 12, 36–7, 53–4, 174, 255–6, 277–8; nomic, 43; real distinguished from epistemic, 42, 256. *See also* natural necessity.
neo-classical equilibrium theory, 184, 196, 226
Newtonianism, 263–5
nominalism: in Locke's theory of kinds, 169; of the program of analysis, 272
non-locality, 218

non-proliferation: principle of, 90, 94, 152–4

objective knowledge: and natural kind structure, xiii
objectivity: in history and social sciences, 180–2
objects or substances: as natural kinds, 22
ontological problem, 11, 215
ontological dependence: in Aristotelian and Platonic theory, 80–82; direction of, 63–5; and ontological reduction, 82
Ontological Regress Argument: for dispositional realism, 115–16
ontology: aims of, 8, 61–2; Armstrong's, 63–5; of facts, 62; of states of affairs, 63–5, 72

Pargetter, Robert J., 24, 66–7, 68–9, 75, 119; and abstract model theories, 152–4; and introverted realism, 148–50
Peirce, C.S., 104n23
Place, U.T., 170–1
Planck's radiation law, 192, 193, 205
Platonism: scientific, 154–7; about universals, 19, 80–1, 83–4, 88, 152–3,155
Poincaré, Henri, 213
Popper, K.R., 40, 188, 276
positivist theories of explanation, 159
possibility: epistemic, 12, 172–3, 180, 220–1, 233, 254–6, 279–80, 289–91; imagined, 221, 231–4, 172–3; logical, 220, 231, 234–7; metaphysical, or real, 53–4, 172–3, 233, 234–7, 280
possible worlds. *See* worlds: possible.
Price, H.H., 18
principles: general structural, 206
Prior, Elizabeth, 115, 119

probability theory: and distinction between reference class and attribute class, 204

processes: causal, 50, 218; energy transmission, 218; natural kinds of, 162–4; stochastic, 50, 250

properties, 61: accidental, 32, 136; active, 95; essential, 136; in Humeanism, 4–5, 44–6; instances defined, 70; intrinsic and extrinsic defined, 27–9, 30; as natural kinds, 22, 23, 25, 86, 97, 99, 163- 4; negative , 95; passive, 95; species determining in biology, 168; structural, 135. *See also* dispositions; categorical properties.

property universals, 3, 18, 68, 89–97

propositions: analytic distinguished from synthetic, 34

quantities: as generic universals, 66, 67; as natural kinds of tropes, 71; and natural properties and relations, 86; as spectral kinds, 83–4, 174

quantum discreteness: as basis of scientific essentialism, 2–3

Quine, W.V.O., 272

Quinton, A., 103n11

Rational Belief Systems (Ellis), 234

realism: causal process, 147, 158; categorical, 110, 114, 115, 128; dispositional, 9, 112, 127–9; empirical, 169; essentialist, 49–50, 245–8; historical, 181–2; introverted, 148–50; possible worlds, 242, 245, 272–4; and properties, 85–6; requirement of in logical analysis, 270–1; standard argument for, 149, 150, 151; scientific entity, 145–6;

strong argument for, 149, 151, 154, 157; transcendental, 169

realistic analysis, 274–8, 279–83, 291–2

Reichenbach, H., 56n6

relations: and causal powers, 136–8; n-adic, 94; quantitative in economics, 193–4

relativism: in social sciences, 182

relativity: principles of, 205

relevant subvenient basis, 82–3

research programs: and hard-core assumptions and protective belts of hypotheses, 184

resemblances: philosophy of, 18

Rosenberg, Alexander, 185, 196

RS-basis. *See* relevant subvenient basis.

Russell, Bertrand, 62, 269

Salmon, W.C., 159

Sankey, Howard, 285

Sawyer, Kim, 191

Schlesinger, George, 126

Schrodinger's equation, 51

semantics: and dispositions, 126–7; possible worlds, 12, 151, 241–2, 248, 274, 282–3; realistic, 112, 274–8, 283

sets, 88–9, 272

Shoemaker, Sidney, 217, 103n12; on dispositions, 138n4, 138n6

singular causation, 268

Smart, J.J.C., 40, 147, 149, 171, 262, 272

social sciences: inapplicability of scientific essentialism to, 10

Special Theory of Relativity, 196

statistical relevance models, 158–9

structural universals: in Bigelow and Pargetter's theories, 68–9, 75

substantive universals, 68, 92, 97, 98–9; defined, 97; distinguished from dynamic and property universals, 3; infimic species of, 98; in hierarchy of natural kinds, 70, 73

supervenience 63–4, 82

Swinburne, Richard, 71, 135

Swinburne's Regress, 71–2, 135–7

Swoyer, Chris, 210, 215, 217

theories: abstract model, 150, 151, 152–4, 183–4; causal process, 149, 150, 157; correspondence, 148–9; of natural necessity, 150

Thompson, R.: and classification of natural kinds, 56n2

Thompson, I.J., 114

Tipler, Frank, 251

Tobin, J., 200n15

Tooley, Michael, 210, 215

transubstantiation, 92, 245–7

tropes, 3, 19, 23–6, 70–1, 74, 80, 88–9, 97–8: defined, 24–5; as instances of property universals, 3, 70, 89

Twin-Earth, 101, 171

universalia in rebus, 82, 83

universals, 67, 258n8; accidental distinguished from nomic, 43, 207, 231; Aristotelian theory of, 19, 80–2; in Bigelow and Pargetter, 154–5; classical and generic, 66, 67, 70, 98; conjunctive, 75, 90–1, 96–7; disjunctive, 90; dynamic , 3, 68, 98–9; and laws of nature, 207; natural kinds as, 97; negative, 101; philosophy of, 18–9; ontology of defined, 18–19; sparse theory of, 18, 82, 89–97; strong interconnectedness thesis, 251–2 *See also* substantive universals; dynamic universals; property universals; hierarchy: of universals; properties.

Van Fraassen, Bas C., 208

vis viva controversy, 263–5, 293–4

Wilkerson, T.E., 170

Williams, Donald, 89

Wittgenstein, L.: ontology of, 62

worlds: alien, 253–5; essential properties of worlds like ours, 249–53; natural kinds of, 249–53; and natural necessity, 214–5; possible, 151, 154, 270–4; sibling, 254. *See also* Hume worlds.

www.ingramcontent.com/pod-product-compliance
Ingram Content Group UK Ltd.
Pitfield, Milton Keynes, MK11 3LW, UK
UKHW040703180125
453697UK00010B/379